Skills Worksheet

Directed Reading A

AF344474

Section: Arranging the Elements

1. Why do you think scientists might have been frustrated by the organization of the elements before 1869?

____

____

____

____

____

DISCOVERING A PATTERN

_______ **2.** Which arrangement of elements did Mendeleev find produced a repeating pattern of properties?
  a. by increasing density
  b. by increasing melting point
  c. by increasing shine
  d. by increasing atomic mass

3. When something occurs or repeats at regular intervals, it is called

_______________________________ .

4. Mendeleev's table, which shows elements' properties following a pattern that

repeats every seven elements, is called the _____________________________ .

5. How was it possible that Mendeleev was able to predict the properties of elements that no one knew about?

____

____

____

____

Copyright © by Holt, Rinehart and Winston. All rights reserved.

CHANGING THE ARRANGEMENT

_______ **6.** How did Moseley solve the problem of the elements that did not fit the pattern according to their properties?
- **a.** He rearranged the elements by atomic mass.
- **b.** He discovered protons, neutrons, and electrons.
- **c.** He disproved the periodic law.
- **d.** He determined the elements' atomic number and then arranged them by atomic number.

7. When the repeating chemical and physical properties of elements change periodically with the elements' atomic numbers, it is called the

_______________________________.

PERIODIC TABLE OF THE ELEMENTS

_______ **8.** Which information is NOT included in each square of the periodic table in your text?
- **a.** atomic number
- **b.** chemical symbol
- **c.** melting point
- **d.** atomic mass

9. How can you tell on the periodic table that carbon is a solid at room temperature?

____

____

____

THE PERIODIC TABLE AND CLASSES OF ELEMENTS

10. Elements are classified as metals, nonmetals, or metalloids according to their

_______________________________.

11. The number of _______________________ in the outer energy level of an atom helps determine which category an element belongs in.

12. How can the zigzag line on the periodic table help you?

____

____

____

Copyright © by Holt, Rinehart and Winston. All rights reserved.

Directed Reading A *continued*

13. Most elements are ____________________, which can be found to the left of the zigzag line on the periodic table.

14. Most metals are ____________________, which means that they can be drawn into thin wires.

15. Most metals are ____________________ at room temperature.

16. Most metals are malleable. What does this mean?

____

____

17. What metal is flattened into sheets that are made into cans and foil?

____

18. What elements are found to the right of the zigzag line on the periodic table?

____

19. Semiconductors, also called ____________________, are the elements that border the zigzag line on the periodic table.

DECODING THE PERIODIC TABLE

______**20.** Which elements often share properties?
    a. those in a period   **c.** those with the same color
    b. those in a group   **d.** those in a horizontal row

______**21.** The physical and chemical properties of the elements change
    a. within a group.   **c.** across each period.
    b. within a family.   **d.** across each group.

22. For most elements, the ____________________ has one or two letters, with the first letter always capitalized.

23. Horizontal rows of elements on the periodic table are called

____________________.

24. Vertical columns of elements on the periodic table are called

____________________, or ____________________.

25. Some elements, such as ____________________, are named after

scientists. Others, such as ____________________, are named after places.

Copyright © by Holt, Rinehart and Winston. All rights reserved.

Directed Reading A

Section: Grouping the Elements

______ **1.** What gives elements in a family or group similar properties?
   a. the same atomic mass
   b. the same number of protons in their nuclei
   c. the same number of electrons in their outer energy level
   d. the same number of total electrons

GROUP 1: ALKALI METALS

______ **2.** Which of the following is NOT true of alkali metals?
   a. They can be cut with a knife.
   b. They are usually stored in water.
   c. They are the most reactive of all the metals.
   d. They can easily give away their outer electron.

3. Metals that share both physical and chemical properties are called

   ________________________ .

GROUP 2: ALKALINE-EARTH METALS

4. Atoms of ____________________ have two outer-level electrons.

5. What are two products made from calcium compounds?

   ____

   ____

   ____

6. In what way does calcium help you?

   ____

   ____

   ____

7. Name three alkaline-earth metals besides calcium.

   ____

   ____

   ____

Copyright © by Holt, Rinehart and Winston. All rights reserved.

Directed Reading A *continued*

GROUPS 3–12: TRANSITION METALS

______ **8.** Which of the following characteristics does NOT describe transition metals?
a. They are good conductors of thermal energy.
b. They are more reactive than alkali and alkaline-earth metals.
c. They have one or two electrons in the outer energy level.
d. They are denser than elements in Groups 1 and 2.

9. Metals that are less reactive than alkali metals and alkaline-earth metals are

called _______________________.

10. How is mercury different from other transition metals?

____

____

11. Two rows of transition metals are placed at the bottom of the periodic table

to save space. Elements in the first row are called _______________________.

Elements in the second row are called _______________________.

12. Which lanthanide forms a compound that enables you to see red on a
computer screen?

____

13. Which actinide is used in some smoke detectors?

____

GROUP 13: BORON GROUP

14. Why did Emperor Napoleon III of France use aluminum dinnerware?

____

____

____

15. What are some of the uses of aluminum?

____

____

____

Copyright © by Holt, Rinehart and Winston. All rights reserved.

❙ Directed Reading A *continued*

GROUP 14: CARBON GROUP

16. The metalloids _________________ and _________________,
both in Group 14, are used to make computer chips.

17. What are three compounds of carbon that are necessary for living things on
Earth?

____

18. The hardest material known is _________________.

19. What are some of the uses of diamond?

____

____

20. What form of carbon is used as a pigment?

____

GROUP 15: NITROGEN GROUP

21. Nitrogen is a _________________ at room temperature.

22. Each element in the Nitrogen Group has _________________ electrons
in the outer level.

23. Nitrogen from the air can react with what element to make ammonia for
fertilizer?

____

GROUP 16: OXYGEN GROUP

24. How is oxygen different from the other four elements in Group 16?

____

____

25. The element _________________ can be found as a yellow solid in
nature and is used to make sulfuric acid.

26. Why is oxygen important?

____

____

____

Copyright © by Holt, Rinehart and Winston. All rights reserved.

Directed Reading A *continued*

GROUP 17: HALOGENS

27. The atoms of _________________________ need to gain only one electron to have a complete outer level.

28. What important use do the halogens iodine and chlorine have in common?

____

____

29. Halogens combine with most metals to form _________________________, such

as _________________________.

30. How does chlorinating water help protect people?

____

____

____

GROUP 18: NOBLE GASES

______**31.** Which of the following statements about noble gases is NOT true?
    a. They are colorless and odorless at room temperature.
    b. They have a complete set of electrons in their outer energy level.
    c. They normally react with other elements.
    d. All of them are found in Earth's atmosphere in small amounts.

32. The atoms of _________________________ have a full set of electrons in their outer level.

33. The low _________________________ of helium makes blimps and weather balloons float.

HYDROGEN

______**34.** Which of the following statements about hydrogen is NOT true?
    a. It is useful as rocket fuel.
    b. It is the most abundant element in the universe.
    c. Its physical properties are closer to those of nonmetals than to those of metals.
    d. It has two electrons in its outer energy level.

Copyright © by Holt, Rinehart and Winston. All rights reserved.

Skills Worksheet

Directed Reading B

Section: Arranging the Elements

<u>Circle the letter</u> of the best answer for each question.

DISCOVERING A PATTERN

1. How did Mendeleev arrange the elements?

a. by increasing density

b. by increasing melting point

c. by increasing shine

d. by increasing atomic mass

Periodic Properties of the Elements

2. What does periodic mean?

a. happening at regular intervals

b. happening almost never

c. happening twice

d. happening once or twice

3. Mendeleev's pattern repeated after how many elements?

a. every two elements

b. every three elements

c. every seven elements

d. every five elements

Predicting Properties of Missing Elements

4. How did Mendeleev predict the properties of missing elements?

a. by making the properties up

b. by comparing the properties of found elements

c. by using the pattern he found

d. by not following the pattern

Copyright © by Holt, Rinehart and Winston. All rights reserved.

Directed Reading B *continued*

CHANGING THE ARRANGEMENT

Circle the letter of the best answer for each question.

5. What does the periodic law say about the properties of elements?

    a. They change with atomic mass.

    b. They change with atomic number.

    c. They change with mass number.

    d. They change when the isotope changes.

6. How many elements discovered since 1914 follow the periodic law?

    a. all of them

    b. about half of them

    c. none of them

    d. every seventh element

PERIODIC TABLE OF THE ELEMENTS

7. What is NOT included in each square of the periodic table in your text?

    a. atomic number

    b. chemical symbol

    c. melting point

    d. atomic mass

8. What color indicates an element is a solid?

    a. red

    b. blue

    c. green

    d. yellow

Copyright © by Holt, Rinehart and Winston. All rights reserved.

Directed Reading B *continued*

THE PERIODIC TABLE AND CLASSES OF ELEMENTS

Read the words in the box. Read the sentences. <u>Fill in each blank</u> with the word or phrase that best completes the sentence.

nonmetal	semiconductors	metals	metalloids

9. Most of the elements in the periodic table are

_______________________ .

10. More than half of the _______________________ are gases at

room temperature.

11. The atoms of _______________________ have about half of a

complete set of electrons in their outer energy level.

12. Metalloids are also called _______________________ .

DECODING THE PERIODIC TABLE

Each Element Is Identified by a Chemical Symbol

mendelevium	chemical symbol	californium

13. Elements such as _______________________ are named after

scientists.

14. Elements such as _______________________ are named after

places.

15. For most elements, the _______________________ has one or

two letters.

Copyright © by Holt, Rinehart and Winston. All rights reserved.

Directed Reading B *continued*

Rows Are Called *Periods*

Circle the letter of the best answer for each question.

16. What is each horizontal row called?

   a. group

   b. family

   c. period

   d. property

17. How do you read periods?

   a. from top to bottom

   b. from bottom to top

   c. from left to right

   d. from right to left

18. How do the physical and chemical properties of the elements change?

   a. within a group

   b. within a family

   c. across each period

   d. across each group

Columns Are Called *Groups*

19. What is another name for *group*?

   a. period

   b. family

   c. element

   d. electron

20. Which elements usually have the same properties?

   a. those in a period

   b. those in a group

   c. those with the same color

   d. those in a horizontal row

Copyright © by Holt, Rinehart and Winston. All rights reserved.

Skills Worksheet

Directed Reading B

Section: Grouping the Elements

<u>Circle the letter</u> of the best answer for each question.

1. What do atoms of elements in a group have that makes their properties similar?

    a. the same atomic mass

    b. the same number of protons

    c. the same number of outer level electrons

    d. the same total number of electrons

2. At the atomic level, what makes elements reactive?

    a. having a filled outer level

    b. exchanging or sharing electrons

    c. having the same number of electrons

    d. having the same number of protons

GROUP 1: ALKALI METALS

3. Look at the chart. What is the symbol for potassium?

    a. Li

    b. K

    c. Cs

    d. Fr

4. How are the alkali metals similar?

    a. They are very reactive.

    b. They have few uses.

    c. They are so hard they cannot be cut.

    d. They are often stored in water.

Copyright © by Holt, Rinehart and Winston. All rights reserved.

Directed Reading B *continued*

GROUP 2: ALKALINE-EARTH METALS
Circle the letter of the best answer for each question.

5. What is true about all of the alkaline-earth metals?

   a. They are have low density.

   b. They are less reactive than alkali metals are.

   c. They have three outer level electrons.

   d. They have few uses.

6. Look at the chart. Which element is not an alkaline-earth metal?

   a. magnesium

   b. calcium

   c. barium

   d. sodium

GROUPS 3–12: TRANSITION METALS

7. Groups are read from top to bottom. What elements are in Group 3?

   a. Sc, Ti, V, Mn

   b. V, Nb, Ta, Db

   c. Sc, Y, La, Ac

   d. La, Hf, Ta, W

Copyright © by Holt, Rinehart and Winston. All rights reserved.

Properties of Transition Metals
Circle the letter of the best answer for each question.

8. How would you describe most transition metals?

   a. poor conductor of electric current

   b. dull

   c. good conductor of thermal energy

   d. low density and melting points

Lanthanides and Actinides

9. What word best describes the lathanides?

   a. radioactive

   b. unstable

   c. reactive

   d. dull

10. What word best describes the actinides?

   a. radioactive

   b. stable

   c. shiny

   d. reactive

Copyright © by Holt, Rinehart and Winston. All rights reserved.

| Directed Reading B *continued*

Read the words in the box. Read the sentences. <u>Fill in each blank</u> with the word or phrase that best completes the sentence.

reactive	oxygen	hydrogen
aluminum	tin	carbohydrates

GROUP 13: BORON GROUP

11. All elements in the Boron Group are _______________________________

and solid at room temperature.

12. The most common element in the Boron Group is

_______________________________ which is used to make airplane parts.

GROUP 14: CARBON GROUP

13. Compounds of carbon, such as proteins, fats, and

_______________________________, are necessary for life on Earth.

14. The symbol Sn stands for the metal _______________________________.

GROUP 15: NITROGEN GROUP

15. Nitrogen can react with _______________________________ to

make ammonia.

GROUP 16: OXYGEN GROUP

16. In order for a substance to burn, it needs

_______________________________.

Copyright © by Holt, Rinehart and Winston. All rights reserved.

Directed Reading B *continued*

GROUP 17: HALOGENS

<u>Circle the letter</u> of the best answer for each question.

17. What is made when a halogen reacts with a metal?

   a. a salt

   b. a compound

   c. a nonmetal

   d. an electron

GROUP 18: NOBLE GASES

18. Look at the chart. How many nonmetals are in Group 18?

   a. three

   b. six

   c. four

   d. eight

19. Look at the chart. What element does the symbol Kr represent?

   a. argon

   b. helium

   c. neon

   d. krypton

20. What element helps light bulbs last longer?

   a. neon

   b. krypton

   c. argon

   d. xenon

Copyright © by Holt, Rinehart and Winston. All rights reserved.

Directed Reading B *continued*

HYDROGEN

<u>**Circle the letter**</u> **of the best answer for each question.**

21. What is the symbol for hydrogen?

    a. Hy

    b. K

    c. H

    d. Hi

22. Which word or words describe hydrogen?

    a. colorless gas

    b. unreactive

    c. high density

    d. very rare

23. Where is hydrogen located on the periodic table?

    a. in Group 1

    b. in Group 18

    c. above Group 1

    d. below Group 1

Copyright © by Holt, Rinehart and Winston. All rights reserved.

Skills Worksheet

Vocabulary and Section Summary

Arranging the Elements

VOCABULARY

In your own words, write a definition of the following terms in the space provided.

1. periodic

____

____

2. periodic law

____

____

3. period

____

____

4. group

____

____

SECTION SUMMARY

Read the following section summary.

- Mendeleev developed the first periodic table by listing the elements in order of increasing atomic mass. He used his table to predict that elements with certain properties would be discovered later.
- Properties of elements repeat in a regular, or periodic, pattern.
- Moseley rearranged the elements in order of increasing atomic number.
- The periodic law states that the repeating chemical and physical properties of elements relate to and depend on elements' atomic numbers.
- Elements in the periodic table are classified as metals, nonmetals, and metalloids.
- Each element has a chemical symbol.
- A horizontal row of elements is called a *period.*
- Physical and chemical properties of elements change across each period.
- A vertical column of elements is called a *group* or *family.*
- Elements in a group usually have similar properties.

Copyright © by Holt, Rinehart and Winston. All rights reserved.

Skills Worksheet

Vocabulary and Section Summary

Grouping the Elements

VOCABULARY

In your own words, write a definition of the following terms in the space provided.

1. alkali metal

____

____

2. alkaline-earth metal

____

____

3. halogen

____

____

4. noble gas

____

____

SECTION SUMMARY

Read the following section summary.

- Alkali metals (Group 1) are the most reactive metals. Atoms of the alkali metals have one electron in their outer level.

- Alkaline-earth metals (Group 2) are less reactive than the alkali metals are. Atoms of the alkaline-earth metals have two electrons in their outer level.

- Transition metals (Groups 3–12) include most of the well-known metals and the lanthanides and actinides.

- Groups 13–16 contain the metalloids and some metals and nonmetals.

- Halogens (Group 17) are very reactive nonmetals. Atoms of the halogens have seven electrons in their outer level.

- Noble gases (Group 18) are unreactive nonmetals. Atoms of the noble gases have a full set of electrons in their outer level.

- Hydrogen is set off by itself. Its properties do not match the properties of any one group.

Copyright © by Holt, Rinehart and Winston. All rights reserved.

Skills Worksheet

Section Review

Arranging the Elements

USING KEY TERMS

1. In your own words, write a definition for the term *periodic*.

____

____

UNDERSTANDING KEY IDEAS

_______ **2.** Which of the following elements should be the best conductor of electric current?

    a. germanium

    b. sulfur

    c. aluminum

    d. helium

3. Compare a period and a group on the periodic table.

____

____

4. What property did Mendeleev use to position the elements on the periodic table?

____

____

5. State the periodic law.

____

____

CRITICAL THINKING

6. Identifying Relationships An atom that has 117 protons in its nucleus has not yet been made. Once this atom is made, to which group will element 117 belong? Explain your answer.

____

____

____

Copyright © by Holt, Rinehart and Winston. All rights reserved.

Section Review *continued*

7. Applying Concepts Are the properties of sodium, Na, more like the properties of lithium, Li, or magnesium, Mg? Explain your answer.

____

____

____

____

INTERPRETING GRAPHICS

8. The image below shows part of a periodic table. Compare the image below with the similar part of the periodic table in your book.

1	1 **H** 1.0079 水素	
2	3 **Li** 6.941 リチウム	4 **Be** 9.01218 ベリリウム
3	11 **Na** 22.98977 ナトリウム	12 **Mg** 24.305 マグネシウム
	19 K	20 Ca	21 Sc	22 Ti

____

____

____

____

Copyright © by Holt, Rinehart and Winston. All rights reserved.

Skills Worksheet

Section Review

Grouping the Elements
USING KEY TERMS

Complete each of the following sentences by choosing the correct term from the word bank.

noble gas	alkaline-earth metal
halogen	alkali metal

1. An atom of a(n) _______________________ has a full set of electrons in its outermost energy level.

2. An atom of a(n) _______________________ has one electron in its outermost energy level.

3. An atom of a(n) _______________________ tends to gain one electron when it combines with another atom.

4. An atom of a(n) _______________________ tends to lose two electrons when it combines with another atom.

UNDERSTANDING KEY IDEAS

______ **5.** Which group contains elements whose atoms have six electrons in their outer level?

    a. Group 2   **c.** Group 16

    b. Group 6   **d.** Group 18

6. What are two properties of the alkali metals?

____

____

7. What causes the properties of elements in a group to be similar?

____

____

8. What are two properties of the halogens?

____

____

9. Why is hydrogen set apart from the other elements in the periodic table?

____

____

Copyright © by Holt, Rinehart and Winston. All rights reserved.

| Section Review *continued*

10. Which group contains elements whose atoms have three electrons in their
outer level?

____

____

INTERPRETING GRAPHICS

11. Look at the model of an atom below. Does the model represent a metal atom
or a nonmetal atom? Explain your answer.

____

____

____

CRITICAL THINKING

12. Making Inferences Why are neither the alkali metals nor the alkaline-earth
metals found uncombined in nature?

____

____

____

13. Making Comparisons Compare the element hydrogen with the alkali metal
sodium.

____

____

____

Copyright © by Holt, Rinehart and Winston. All rights reserved.

Chapter Review

USING KEY TERMS

Complete each of the following sentences by choosing the correct term from the word bank.

group	period	alkali metals
halogens	alkaline-earth metals	noble gases

1. Elements in the same vertical column on the periodic table belong to the

same _____________________.

2. Elements in the same horizontal row on the periodic table belong to the

same _____________________.

3. The most reactive metals are _____________________.

4. Elements that are unreactive are called _____________________.

UNDERSTANDING KEY IDEAS

Multiple Choice

_______ **5.** Mendeleev's periodic table was useful because it
  a. showed the elements arranged by atomic number.
  b. had no empty spaces.
  c. showed the atomic number of the elements.
  d. allowed for the prediction of the properties of missing elements.

_______ **6.** Most nonmetals are
  a. shiny.
  b. poor conductors of electric current.
  c. flattened when hit with a hammer.
  d. solids at room temperature.

_______ **7.** Which of the following items is NOT found on the periodic table?
  a. the atomic number of each element
  b. the name of each element
  c. the date that each element was discovered
  d. the atomic mass of each element

_______ **8.** Which of the following statements about the periodic table is false?
  a. There are more metals than nonmetals on the periodic table.
  b. Atoms of elements in the same group have the same number of electrons in their outer level.
  c. The elements at the far left of the periodic table are nonmetals.
  d. Elements are arranged by increasing atomic number.

Copyright © by Holt, Rinehart and Winston. All rights reserved.

| Chapter Review *continued*

______ **9.** Which of the following statements about alkali metals is true?
  a. Alkali metals are generally found in their uncombined form.
  b. Alkali metals are Group 1 elements.
  c. Alkali metals should be stored underwater.
  d. Alkali metals are unreactive.

______ **10.** Which of the following statements about elements is true?
  a. Every element occurs naturally.
  b. All elements are found in their uncombined form in nature.
  c. Each element has a unique atomic number.
  d. All of the elements exist in approximately equal quantities.

Short Answer

11. How is Moseley's basis for arranging the elements different from Mendeleev's?

____

____

12. How is the periodic table like a calendar?

____

____

Math Skills

Examine the chart of the percentages of elements in the Earth's crust below. Then, answer the questions that follow.

13. Excluding the "Other" category, what percentage of the Earth's crust are alkali metals?

____

14. Excluding the "Other" category, what percentage of the Earth's crust are alkaline-earth metals?

____

Copyright © by Holt, Rinehart and Winston. All rights reserved.

| Chapter Review *continued*

CRITICAL THINKING

15. Concept Mapping Use the following terms to create a concept map:
periodic table, elements, groups, periods, metals, nonmetals, and *metalloids.*

Copyright © by Holt, Rinehart and Winston. All rights reserved.

| Chapter Review *continued*

16. Forming Hypotheses Why was Mendeleev unable to make any predictions about the noble gas elements?

____

____

____

17. Identifying Relationships When an element that has 115 protons in its nucleus is synthesized, will it be a metal, a nonmetal, or a metalloid? Explain your answer.

____

____

____

18. Applying Concepts Your classmate offers to give you a piece of sodium that he found on a hiking trip. What is your response? Explain.

____

____

____

19. Applying Concepts Identify each element described below.

a. This metal is very reactive, has properties similar to those of magnesium, and is in the same period as bromine.

____

____

____

b. This nonmetal is in the same group as lead.

____

____

____

Copyright © by Holt, Rinehart and Winston. All rights reserved.

| Chapter Review *continued*

INTERPRETING GRAPHICS

20. Study the diagram below to determine the pattern of the images. Predict the missing image, and draw it. Identify which properties are periodic and which properties are shared within a group.

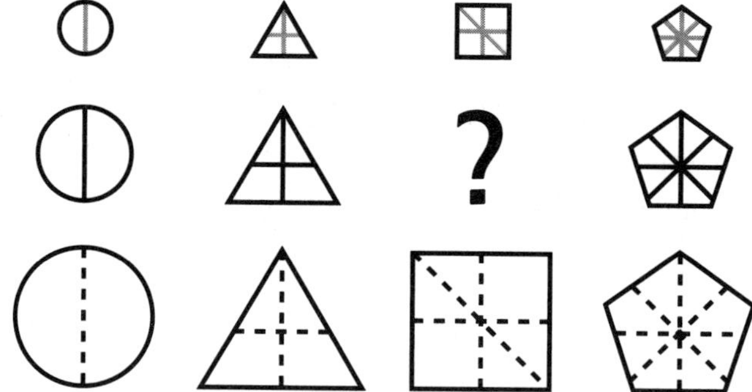

____

____

____

____

Copyright © by Holt, Rinehart and Winston. All rights reserved.

Skills Worksheet)

Reinforcement

Placing All Your Elements on the Table

Complete this worksheet after you have finished reading the section "Grouping the Elements."

You can tell a lot about the properties of an element just by looking at the element's location on the periodic table. This worksheet will help you better understand the connection between the periodic table and the properties of the elements. Follow the directions below, and use crayons or colored pencils to color the periodic table at the bottom of the page.

1. Color the square for hydrogen yellow.

2. Color the groups with very reactive metals red.

3. Color and label the noble gases orange.

4. Color the transition metals green.

5. Using black, mark the zigzag line that shows the position of the metalloids.

6. Color the metalloids purple.

7. Use blue to color all of the non-metals that are not noble gases.

8. Color the metals in Groups 13–16 brown.

9. Circle and label the actinides in yellow.

10. Circle and label the lanthanides in red.

11. Circle and label the alkali metals in blue.

12. Circle and label the alkaline-earth metals in purple.

13. Circle and label the halogens in green.

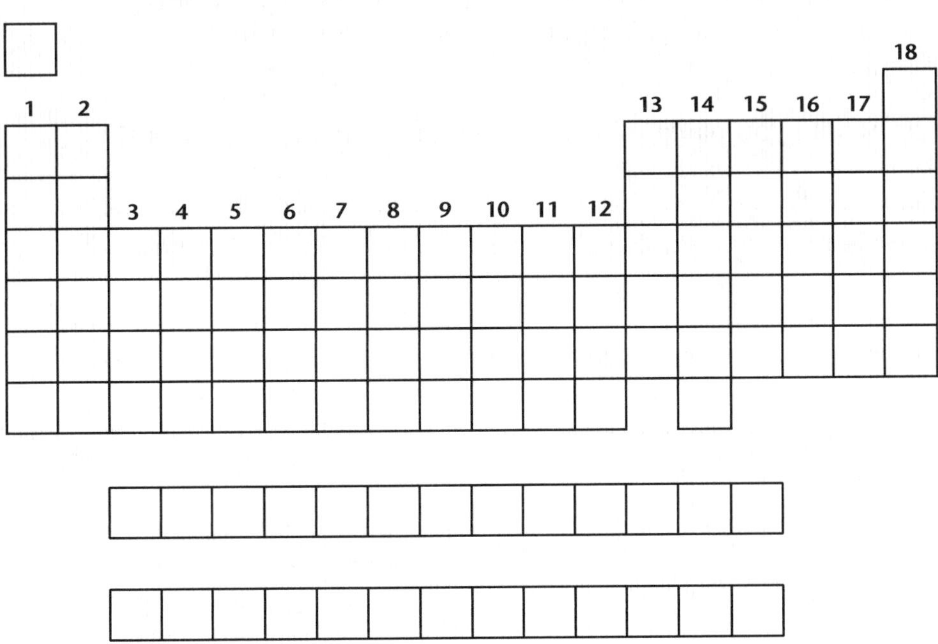

Copyright © by Holt, Rinehart and Winston. All rights reserved.

Reinforcement *continued*

Answer the following questions using the periodic table on the previous page.

14. The alkaline-earth metals react similarly because they all have the same number of electrons in their outer energy level. Which group contains the alkaline-earth metals?

____

15. How many electrons are in the outer energy level of the atoms of alkaline-earth metals?

____

16. Hydrogen is in a different color than the rest of the elements in Group 1. Give an example of how hydrogen's characteristics set it apart from other Group 1 elements.

____

____

17. What is the name of the group of unreactive nonmetals that includes argon?

____

18. Except for the metalloids, what do all of the elements on the right side of the zigzag line have in common?

____

____

Imagine you are a scientist who has just discovered a new element. The element has an atomic number of 113, and it has three electrons in the outer energy level of each atom.

19. Where would you place this new element in the periodic table?

____

20. Which element would have properties most similar to the new element?

____

21. What name would you suggest for this new element?

____

Copyright © by Holt, Rinehart and Winston. All rights reserved.

Name _______________________________ Class _______________ Date _______________

Critical Thinking

A Solar Solution

While searching the Internet for new science products, you come across a bulletin board advertising the following items. Use your knowledge of the periodic table of elements to review the following advertisement for accuracy:

Acme Science Products

- **NEW AND IMPROVED "ACME SALT"** —100% sodium. Because it is found in nature, it is 100% PURE.

- **NEW!** Experimental electrical wire made entirely of sulfur. Get yours while supplies last.

- **ELIMINATE WATER BILLS** by using the new "Acme Thirst Buster 2" water system. With an electric spark, it combines oxygen and hydrogen to create your own water supply at home.

- The Acme **"EVERLAST LIGHT BULB"** will burn twice as long as other bulbs because it is filled with oxygen.

- Acme has discovered **A BRAND NEW ELEMENT.** Find out more on our home page!"

EVALUATING INFORMATION

1. What is wrong with the Acme salt ad?

____

____

____

____

DEMONSTRATING REASONED JUDGMENT

2. Would buying sulfur electric wire be a wise choice? Explain.

____

____

____

____

____

Copyright © by Holt, Rinehart and Winston. All rights reserved.

| Critical Thinking *continued*

PREDICTING CONSEQUENCES

3. Do you think that using electricity to combine oxygen and hydrogen in your home could cause a problem?

____

____

____

____

4. Would the Acme "Everlast Light Bulb" last longer than an ordinary bulb? Explain.

____

____

____

____

COMPREHENDING IDEAS

5. Acme claims to have discovered a new element. How can you determine if this claim is true?

____

____

____

____

6. How would you go about classifying this new element?

____

____

____

____

Copyright © by Holt, Rinehart and Winston. All rights reserved.

Assessment

Section Quiz

Section: Arranging the Elements

Write the letter of the correct answer in the space provided.

_______ **1.** *Periodic* means
   a. happening at regular intervals.
   b. happening very rarely.
   c. happening frequently.
   d. happening three or four times a year.

_______ **2.** Periodic law states that
   a. elements are either gases, solids, or liquids.
   b. mercury is a liquid at room temperature.
   c. properties of elements change periodically with the elements'
   atomic numbers.
   d. some elements only stay in a liquid state for short periods.

_______ **3.** Each vertical column on the periodic table is called a(an)
   a. period.
   b. group.
   c. element.
   d. property.

_______ **4.** The elements to the right of the zigzag line on the period table are
   called
   a. nonmetals.
   b. metals.
   c. metalloids.
   d. conductors.

_______ **5.** Most metals are
   a. solid at room temperature.
   b. bad conductors of electric current.
   c. dull.
   d. not malleable.

Copyright © by Holt, Rinehart and Winston. All rights reserved.

Section Quiz

Section: Grouping the Elements

Match the correct description with the correct term. Write the letter in the space provided.

_______ **1.** metals that are so reactive that in nature they are found only combined with other elements

_______ **2.** metals that have two outer-level electrons

_______ **3.** shiny, reactive metals, some of which are used to make steel

_______ **4.** elements whose atoms are radioactive

_______ **5.** very reactive nonmetals

_______ **6.** unreactive nonmetals that do not react with other elements under normal conditions

_______ **7.** metals in Groups 3–12 that do not give away their electrons as easily as atoms of Groups 1 and 2

_______ **8.** the most common element in Group 13, the Boron Group

_______ **9.** nonmetal that forms a wide variety of compounds, such as proteins, fats, and carbohydrates

_______ **10.** an element that is necessary for substances to burn

a. aluminum

b. transition metals

c. lanthanides

d. oxygen

e. actinides

f. carbon

g. halogens

h. alkali metals

i. alkaline-earth metals

j. noble gases

Copyright © by Holt, Rinehart and Winston. All rights reserved.

Chapter Test A

The Periodic Table
MULTIPLE CHOICE
Write the letter of the correct answer in the space provided.

_______ **1.** Most of the elements in the periodic table are
   a. metals.
   b. metalloids.
   c. gases.
   d. nonmetals.

_______ **2.** Mendeleev arranged the elements by
   a. density.
   b. melting point.
   c. appearance.
   d. increasing atomic mass.

_______ **3.** The horizontal row on the period table is called a(n)
   a. group.
   b. family.
   c. period.
   d. atomic number.

_______ **4.** Which one of the following tells the physical state of an element at
   room temperature?
   a. the atomic number
   b. the color of the chemical symbol
   c. the atomic mass
   d. the element name

_______ **5.** How do the physical and chemical properties of the elements change?
   a. within a group
   b. across each period
   c. within a family
   d. across each group

_______ **6.** What is necessary for substances to burn?
   a. hydrogen
   b. oxygen
   c. helium
   d. carbon

Copyright © by Holt, Rinehart and Winston. All rights reserved.

Chapter Test A *continued*

_______ **7.** Transition metals are
    a. good conductors of thermal energy.
    b. more reactive than alkali metals.
    c. not good conductors of electric current.
    d. used to make aluminum.

_______ **8.** The elements' properties follow a pattern that repeats every
    a. 7 elements.
    b. 5 elements.
    c. 14 elements.
    d. 10 elements.

_______ **9.** The vertical column of elements on the periodic table is called a(n)
    a. period.
    b. semiconductor.
    c. atomic mass.
    d. group.

MATCHING

Match the correct description with the correct term. Write the letter in the space provided.

_______ **10.** These metals react with water to form hydrogen.

_______ **11.** This metal, part of the Boron Group, is used for aircraft parts.

_______ **12.** This is important to most living things.

_______ **13.** Cement and chalk are compounds of this.

_______ **14.** Some of these reactive metals are used to make steel.

_______ **15.** Chlorine and iodine are these.

_______ **16.** This is a colorless, odorless gas.

_______ **17.** Diamond and soot are forms of this.

_______ **18.** This makes up about 80% of the air we breathe.

_______ **19.** Light bulbs last longer when they are filled with this gas.

a. aluminum
b. argon
c. halogens
d. nitrogen
e. oxygen
f. alkali metals
g. calcium
h. carbon
i. hydrogen
j. lanthanides

Copyright © by Holt, Rinehart and Winston. All rights reserved.

| Chapter Test A *continued*

MULTIPLE CHOICE

Use the figure below to answer questions 20–22.

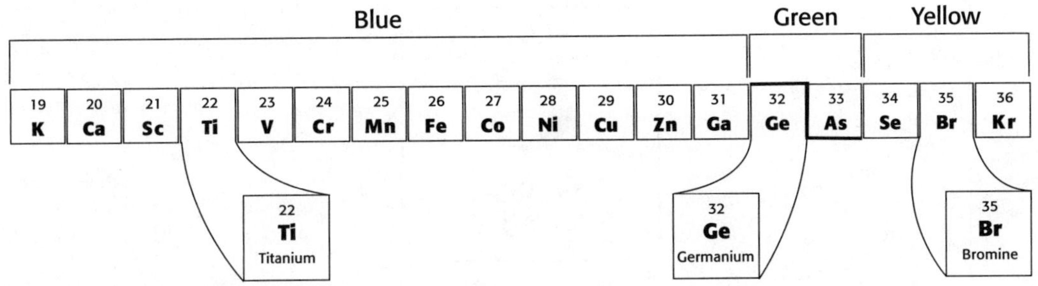

______ **20.** Which of the following elements is the most metallic?
    a. K
    b. Kr
    c. Fe
    d. Cu

______ **21.** Which of these elements is the least metallic?
    a. V
    b. Zn
    c. Co
    d. Se

______ **22.** Which element group has all nonmetals?
    a. K, Ca, Sc
    b. Se, Br, Kr
    c. V, Cr, Mn
    d. As, Se, Br

Copyright © by Holt, Rinehart and Winston. All rights reserved.

MULTIPLE CHOICE

Refer to the figure below to answer questions 23 and 24.

________**23.** The number beneath carbon indicates the
  a. atomic number.
  b. atomic mass.
  c. chemical symbol.
  d. element name.

________**24.** The number at the top is the
  a. atomic number.
  b. element name.
  c. atomic mass.
  d. chemical symbol.

Copyright © by Holt, Rinehart and Winston. All rights reserved.

Assessment

Chapter Test B

The Periodic Table
USING KEY TERMS

Use the terms from the following list to complete the sentences below. Each term may be used only once. Some terms may not be used.

halogens	periodic law	period
periodic	group	noble gases
alkali metals	alkaline-earth metals	actinide

1. The days of the week are _____________________ because they repeat in the same order every seven days.

2. A rule that states that repeating chemical and physical properties of elements change periodically with the atomic number of the elements is

the _____________________.

3. Iodine and chlorine are examples of _____________________.

4. Pure _____________________ are often stored in oil to keep them from reacting with water and oxygen.

5. Atoms of _____________________ have two outer-level electrons.

6. Each up-and-down column of elements on the periodic table is called a(n)

_____________________.

UNDERSTANDING KEY IDEAS

Write the letter of the correct answer in the space provided.

______ **7.** What element makes up about 20% of the air we breathe?
  a. nitrogen
  b. bromine
  c. oxygen
  d. sulfur

______ **8.** Mendeleev found that the elements' properties followed a pattern that repeated every
  a. 7 elements.
  b. 5 elements.
  c. 14 elements.
  d. 10 elements.

______ **9.** The groups of elements that do not have individual names are called the
  a. transition metals.
  b. alkali metals.
  c. alkaline-earth metals.
  d. nonmetals.

Copyright © by Holt, Rinehart and Winston. All rights reserved.

Chapter Test B *continued*

_______**10.** The carbon group has two metalloids, both of which are used to make
    a. dinnerware.   **c.** cans.
    b. foil.   **d.** computer chips.

_______**11.** Diamond and soot are very different, yet both are natural forms of
    a. carbon.   **c.** boron.
    b. nickel.   **d.** copper.

_______**12.** What element is used to make the most widely used compound in the
chemical industry?
    a. sulfur   **c.** selenium
    b. tellurium   **d.** polonium

13. State the periodic law, which is the basis for the periodic table.

____

____

____

____

____

____

14. Explain why hydrogen is unique.

____

____

____

____

____

15. What generalizations can you make about transition metals?

____

____

____

____

Copyright © by Holt, Rinehart and Winston. All rights reserved.

| Chapter Test B *continued*

CRITICAL THINKING

16. Compare the lanthanides and the actinides.

____

____

____

____

____

____

____

17. Carbon forms many important compounds. Could life exist without carbon? Explain your answer.

____

____

____

____

____

____

____

18. How does the unreactivity of noble gases make them useful?

____

____

____

____

____

Copyright © by Holt, Rinehart and Winston. All rights reserved.

Chapter Test B *continued*

CONCEPT MAPPING

19. Use the following terms to complete the concept map below:

nonmetals	metals	solids
gases	shiny	metalloids

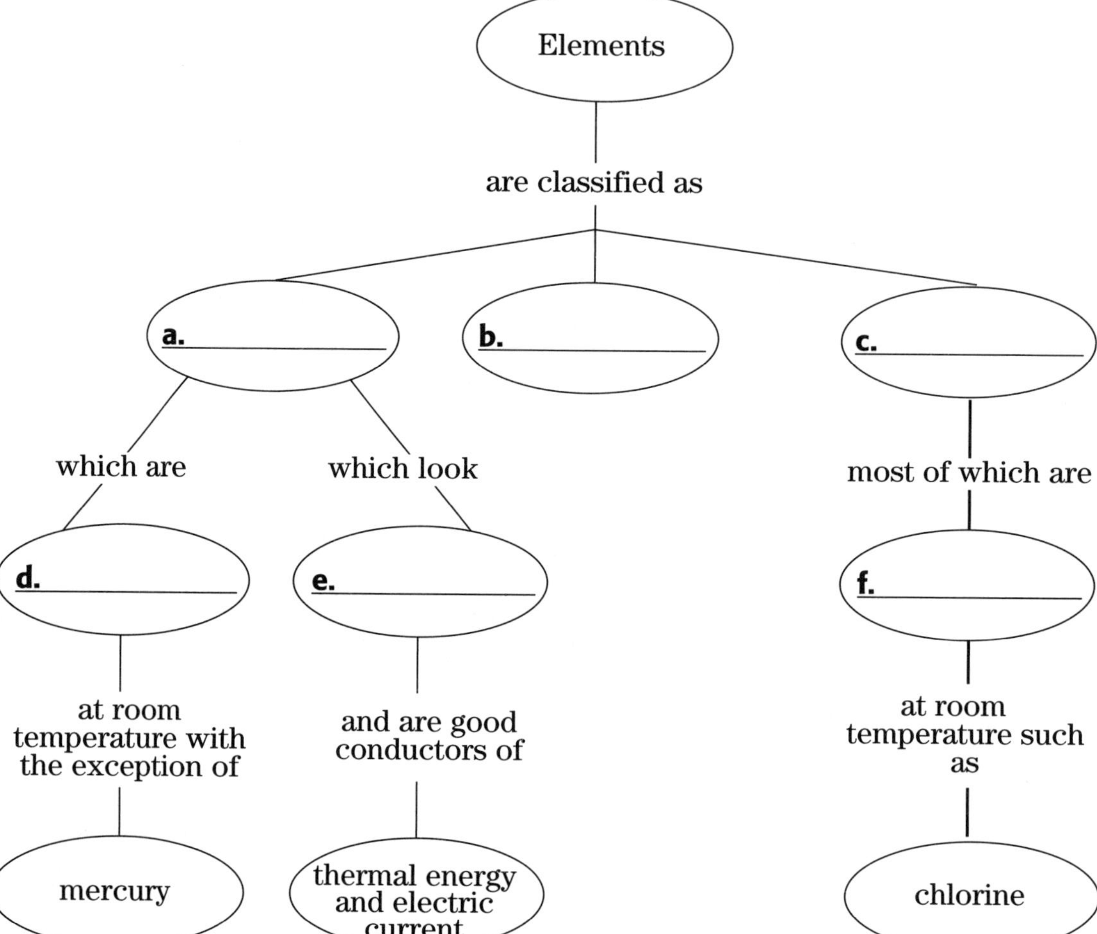

Copyright © by Holt, Rinehart and Winston. All rights reserved.

Chapter Test C

The Periodic Table
MULTIPLE CHOICE
<u>Circle the letter</u> **of the best answer for each question.**

1. What are most of the elements in the periodic table?

   a. metals

   b. metalloids

   c. precious metals

   d. nonmentals

2. How did Mendeleev group the elements?

   a. by density

   b. by melting point

   c. by appearance

   d. by increasing atomic mass

3. Mendeleev's pattern repeated after how many elements?

   a. every seven elements

   b. every three elements

   c. every five elements

   d. every two elements

4. How would you describe most metals?

   a. They are easily shattered.

   b. They are bad conductors of electric current.

   c. They are dull.

   d. They can be drawn into thin wires.

Copyright © by Holt, Rinehart and Winston. All rights reserved.

Circle the letter of the best answer for each question.

5. How many of the recently discovered elements follow periodic law?

   a. none of them

   c. all of them

   b. every eighth element

   d. half of them

6. What are the left-to-right rows on the periodic table?

   a. periods

   b. families

   c. properties

   d. groups

7. Which of the following is a property of alkali metals?

   a. They are so hard they cannot be cut.

   b. They are very reactive.

   c. They are stored in water.

   d. They have few uses.

8. When a halogen reacts with a metal, what is formed?

   a. a salt

   b. a compound

   c. a nonmetal

   d. an electron

9. What helps light bulbs last longer?

   a. krypton

   b. xenon

   c. argon

   d. neon

Copyright © by Holt, Rinehart and Winston. All rights reserved.

| Chapter Test C *continued*

MATCHING
Read the description. Then, <u>draw a line</u> from the dot to the matching word.

10. group made up of six nonmetals ●   **a.** semiconductor

11. another name for metalloid ●   **b.** mendelevium

12. element named after a scientist ●   **c.** noble gases

13. element named after a state ●   **d.** californium

---

14. word meaning "group" ●   **a.** radioactive

15. word describing actinides ●   **b.** shiny

16. word describing hydrogen ●   **c.** colorless

17. word describing most transition ●   **d.** shiny
metals

Copyright © by Holt, Rinehart and Winston. All rights reserved.

Chapter Test C *continued*

FILL-IN-THE-BLANK

Read the words in the box. Read the sentences. <u>Fill in each blank</u> with the word or phrase that best completes the sentence.

aluminum	silicon	oxygen
carbohydrates	hydrogen	

18. In the Boron Group, the most common element

is ________________________.

19. Proteins, fats, and ______________________, which are

compounds of carbon, are necessary for life on Earth.

20. Nitrogen and ____________________ can be combined to

make ammonia.

21. Germanium and ____________________ are used to make

computer chips.

22. A substance needs ____________________ to burn.

Copyright © by Holt, Rinehart and Winston. All rights reserved.

Performance-Based Assessment

OBJECTIVE

You've read about the periodic table. Now you will have a chance to observe the chemical properties of elements in action! In this activity you will cause a reaction between the iron in steel wool and the oxygen in the air.

KNOW THE SCORE!

As you work through the activity, keep in mind that you will be earning a grade for the following:

- how you work with materials and equipment (30%)
- the quality and clarity of your observations (40%)
- how you use the periodic table to explain those observations (30%)

Using Scientific Methods

ASK A QUESTION

What will happen when iron in steel wood reacts with the air?

MATERIALS AND EQUIPMENT

- graduated cylinder
- protective gloves
- vinegar
- small bowl

- fine steel wool, 0000 grade
- watch or clock
- thermometer
- rubber band

SAFETY INFORMATION 

- Do not touch broken thermometers.
- Wear safety goggles.

FORM A HYPOTHESIS

1. Describe the steel wool before the reaction.

____

____

TEST THE HYPOTHESIS

2. Use the graduated cylinder to measure 50 mL of vinegar. Pour the vinegar into the bowl.

3. Place the steel wool in the vinegar and leave it there for 2 minutes.

Copyright © by Holt, Rinehart and Winston. All rights reserved.

▌Performance-Based Assessment *continued*

4. With a gloved hand, remove the steel wool from the vinegar.

5. Wrap the steel wool around the thermometer. Use a rubber band to hold the steel wool in place.

6. Record the thermometer's starting temperature.

____

7. Keep the steel wool around the thermometer for 10 minutes.

8. Record the thermometer's ending temperature.

____

ANALYZE THE RESULTS

1. Describe the steel wool after the reaction in Step 3.

____

____

2. Record the thermometer's starting temperature.

____

3. Record the thermometer's ending temperature.

____

4. List the two changes that you observed or measured after the reaction.

____

5. Use the locations of iron and oxygen in the periodic table to explain the change in the steel wool's appearance.

____

____

DRAW CONCLUSIONS

6. What would you expect to see if you put steel wool in a chamber filled with chlorine gas? Explain.

____

____

____

Copyright © by Holt, Rinehart and Winston. All rights reserved.

▊ Performance-Based Assessment *continued*

7. What would you expect to see if you put the steel wool in a chamber filled with neon gas? Explain.

____

____

____

____

8. Iron is a transition metal. List two common properties of transition metals.

____

____

____

____

Copyright © by Holt, Rinehart and Winston. All rights reserved.

Standardized Test Preparation

READING

Read each of the passages below. Then, answer the questions that follow each passage.

Passage 1 Napoleon III (1808–1873) ruled as emperor of France from 1852 to 1870. Napoleon III was the nephew of the famous French military leader and emperor Napoleon I. Early in his reign, Napoleon III was an <u>authoritarian</u> ruler. France's economy did well under his dictatorial rule, so the French rebuilt cities and built railways. During the 1850s and 1860s, Napoleon III used aluminum dinnerware because aluminum was more valuable than gold. Despite his wealth and French economic prosperity, Napoleon III lost public support and popularity. So, in 1860, he began a series of reforms that allowed more individual freedoms in France.

_______ **1.** What is the meaning of the word *authoritarian* in the passage?
   A controlling people's thoughts and actions
   B writing books and stories
   C being an expert on a subject
   D being very wealthy

_______ **2.** Which of the following statements best describes why Napoleon III probably changed the way he ruled France?
   F He was getting old.
   G He was unpopular and had lost public support.
   H He had built as many railroads as he could.
   I He used aluminum dinnerware.

_______ **3.** According to the passage, in what year did Napoleon III die?
   A 1808
   B 1873
   C 1860
   D 1852

Copyright © by Holt, Rinehart and Winston. All rights reserved.

Standardized Test Preparation *continued*

Passage 2 Named after architect Buckminster Fuller, buckyballs resemble the geodesic domes that are characteristic of the architect's work. Excitement over buckyballs began in 1985, when scientists projected light from a laser onto a piece of graphite. In the soot that remained, researchers found a completely new kind of molecule! Buckyballs are also found in the soot from a candle flame. Some scientists claim to have detected buckyballs in space. In fact, one suggestion is that buckyballs are at the center of the condensing clouds of gas, dust, and debris that form galaxies.

_______ **1.** Which of the following statements correctly describes buckyballs?
  A They are a kind of dome-shaped building.
  B They are shot from lasers.
  C They were unknown before 1985.
  D They are named for the scientist who discovered them.

_______ **2.** Based on the passage, which of the following statements is an opinion?
  F Buckyballs might be in the clouds that form galaxies.
  G Buckyballs are named after an architect.
  H Scientists found buckyballs in soot.
  I Buckyballs are a kind of molecule.

_______ **3.** According to the passage, why were scientists excited?
  A Buckyballs were found in space.
  B An architect created a building that resembled a molecule.
  C Buckyballs were found to be in condensing clouds of gas that form galaxies.
  D A new kind of molecule was found.

Copyright © by Holt, Rinehart and Winston. All rights reserved.

| Standardized Test Preparation *continued*

INTERPRETING GRAPHICS

Use the image of the periodic table below to answer the questions that follow.

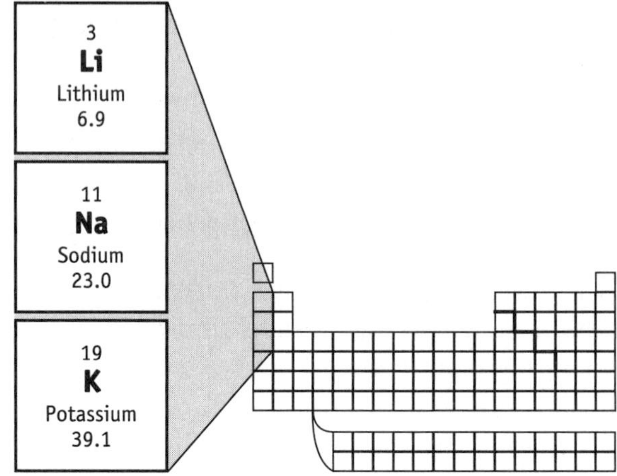

______ **1.** Which of the following statements is correct for the elements shown?
  A Lithium has the greatest atomic number.
  B Sodium has the least atomic mass.
  C Atomic number decreases as you move down the column.
  D Atomic mass increases as you move down the column.

______ **2.** Which of the following statements best describes the outer electrons
  in atoms of the elements shown?
  F The atoms of each element have 1 outer-level electron.
  G Lithium atoms have 3 outer-level electrons, sodium atoms have 11,
  and potassium atoms have 19.
  H Lithium atoms have 7 outer-level electrons, sodium atoms have 23,
  and potassium atoms have 39.
  I The atoms of each element have 11 outer-level electrons.

______ **3.** The elements featured in the image belong to which of the
  following groups?
  A noble gases
  B alkaline-earth metals
  C halogens
  D alkali metals

Copyright © by Holt, Rinehart and Winston. All rights reserved.

Standardized Test Preparation *continued*

MATH

Read each question below, and choose the best answer.

_______ **1.** Elvira's house is 7.3 km from her school. What is this distance expressed in meters?
   A 0.73 m
   B 73 m
   C 730 m
   D 7,300 m

_______ **2.** A chemical company is preparing a shipment of 10 g each of four elements. Each element must be shipped in its own container that is completely filled with the element. Which container will be the largest?

Element	Density (g/cm^3)	Mass (g)
Aluminum	2.702	10
Arsenic	5.727	10
Germanium	5.350	10
Silicon	2.420	10

   F the container of aluminum
   G the container of arsenic
   H the container of germanium
   I the container of silicon

_______ **3.** Arjay has samples of several common elements. Each element has a unique atomic mass (expressed in amu). Which of the following lists shows the atomic masses in order from least to greatest?
   A 63.55, 58.69, 55.85, 58.93
   B 63.55, 58.93, 58.69, 55.85
   C 55.85, 58.69, 58.93, 63.55
   D 55.85, 63.55, 58.69, 58.93

Copyright © by Holt, Rinehart and Winston. All rights reserved.

Model-Making)

DATASHEET FOR CHAPTER LAB

Create a Periodic Table

You probably have classification systems for many things in your life, such as your clothes, your books, and your CDs. One of the most important classification systems in science is the periodic table of the elements. In this lab, you will develop your own classification system for a collection of ordinary objects. You will analyze trends in your system and compare your system with the periodic table of the elements.

OBJECTIVES

Classify objects based on their properties.

Identify patterns and trends in data.

MATERIALS

- bag of objects
- balance, metric
- meterstick
- paper, graphing (2 sheets)
- paper, 3 × 3 cm squares (20)

PROCEDURE

1. Your teacher will give you a bag of objects. Your bag is missing one item. Examine the items carefully. Describe the missing object in as many ways as you can. Be sure to include the reasons why you think the missing object has the characteristics you describe.

____

____

2. Lay the paper squares out on your desk or table so that you have a grid of five rows of four squares each.

3. Arrange your objects on the grid in a logical order. (You must decide what order is logical!) You should end up with one blank square for the missing object.

4. Record a description of the basis for your arrangement.

____

____

5. Measure the mass (g) and diameter (mm) of each object, and record your results in the appropriate square. Each square (except the empty one) should have one object and two written measurements on it.

Copyright © by Holt, Rinehart and Winston. All rights reserved.

Create a Periodic Table *continued*

6. Examine your pattern again. Does the order in which your objects are arranged still make sense? Explain.

____

____

7. Rearrange the squares and their objects if necessary to improve your arrangement. Record a description of the basis for the new arrangement.

____

____

8. Working across the rows, number the squares 1 to 20. When you get to the end of a row, continue numbering in the first square of the next row.

9. Copy your grid below. In each square, be sure to list the type of object and label all measurements with appropriate units.

ANALYZE THE RESULTS

1. **Constructing Graphs** On a separate piece of paper make a graph of mass (y-axis) versus object number (x-axis). Label each axis, and title the graph.

2. **Constructing Graphs** Now make a graph of diameter (y-axis) versus object number (x-axis).

Copyright © by Holt, Rinehart and Winston. All rights reserved.

Create a Periodic Table *continued*

DRAW CONCLUSIONS

3. **Analyzing Graphs** Discuss each graph with your classmates. Try to identify any important features of the graph. For example, does the graph form a line or a curve? Is there anything unusual about the graph? What do these features tell you? Record your answers.

____

____

____

____

____

4. **Evaluating Models** How is your arrangement of objects similar to the periodic table of the elements found in this textbook? How is your arrangement different from that periodic table?

____

____

____

____

5. **Making Predictions** Look again at your prediction about the missing object. Do you think your prediction is still accurate? Try to improve your description by estimating the mass and diameter of the missing object. Record your estimates.

____

____

____

____

6. **Evaluating Methods** Mendeleev created a periodic table of elements and predicted characteristics of missing elements. How is your experiment similar to Mendeleev's work?

____

____

____

Copyright © by Holt, Rinehart and Winston. All rights reserved.

Quick Lab

Conduction Connection

MATERIALS

- cup, plastic-foam
- graphite, mechanical pencil lead
- water, hot
- wire, copper, bare

SAFETY INFORMATION 

PROCEDURE

1. Fill a **plastic-foam cup** with **hot water**.

2. Stand a **piece of copper wire** and a **graphite lead** from a mechanical pencil in the water.

3. After 1 min, touch the top of each object. Record your observations.

____

____

____

4. Which material conducted thermal energy the best? Why?

____

____

____

Copyright © by Holt, Rinehart and Winston. All rights reserved.

Activity

Vocabulary Activity

Bringing It to the Periodic Table

After you finish reading the chapter, try the following puzzle.

On the next page is a partially filled-in quotation by Dmitri Mendeleev. Fill in the term described by each clue below. Then put the numbered letters into the corresponding squares on the next page to find out what Mendeleev said. The answers to questions 9–11 are chemical symbols.

1. states that the properties of elements are periodic functions of their atomic numbers

___ ___ ___ ___ ___ ___ ___ ___   ___ ___ ___
59   16   27 40 24   41

2. column or family in the periodic table

___ ___ ___ ___ ___
19 35 58

3. any element in Groups 3–12

___ ___ ___ ___ ___ ___ ___ ___   ___ ___ ___ ___ ___
31   14   43   55 18   7 33 10

4. elements in Group 1

___ ___ ___ ___ ___ ___   ___ ___ ___ ___ ___ ___
17   22   48   8   36   11

5. having a regular, repeating pattern

___ ___ ___ ___ ___ ___ ___
52   15   25   28 23

6. metals with two electrons in the outer energy level

___ ___ ___ ___ ___ ___ ___ ___-___ ___ ___ ___ ___
51   50   20 42   54   2

7. a row of elements

___ ___ ___ ___ ___ ___
61 6 26   56

8. elements that don't react readily with other elements

___ ___ ___ ___ ___ ___ ___ ___ ___
29   49   62 44   64

9. atomic number 9

___
13

Copyright © by Holt, Rinehart and Winston. All rights reserved.

Vocabulary Activity *continued*

10. atomic number 39

___
57

11. atomic number 54

___ ___
47   63

12. elements having properties of metals and nonmetals

___ ___ ___ ___ ___ ___ ___ ___ ___
39   46   37   5   12

13. the first rows of transition metals at the bottom of the periodic table

___ ___ ___ ___ ___ ___ ___ ___ ___ ___
          1   9   34   4

14. the most abundant element in the universe

___ ___ ___ ___ ___ ___ ___ ___
          21   38   3

15. group containing iodine and chlorine

___ ___ ___ ___ ___ ___ ___ ___
32   60   30   53   45

MENDELEEV'S QUOTATION:

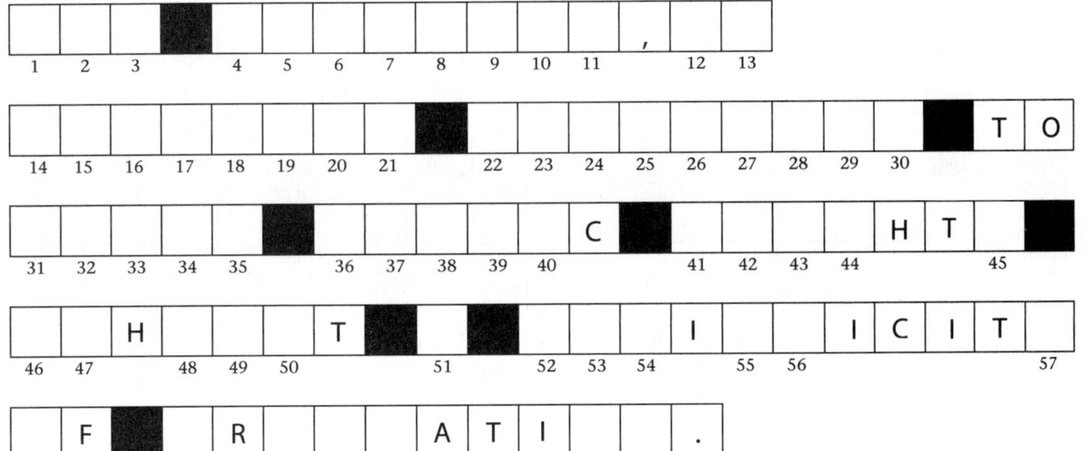

Copyright © by Holt, Rinehart and Winston. All rights reserved.

SciLinks Activity

THE PERIODIC TABLE

Go to www.scilinks.org. To find links related to the periodic table, type in the keyword HSM1125. Then, use the links to answer the following questions about the periodic table.

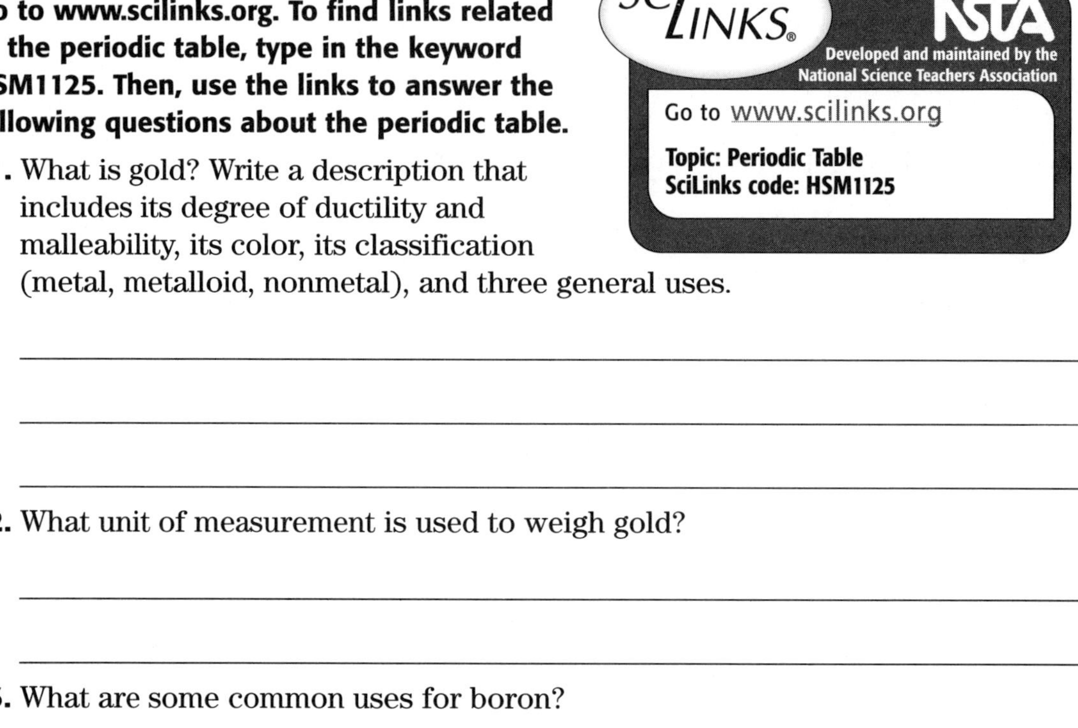

1. What is gold? Write a description that includes its degree of ductility and malleability, its color, its classification (metal, metalloid, nonmetal), and three general uses.

____

____

____

2. What unit of measurement is used to weigh gold?

____

____

3. What are some common uses for boron?

____

____

____

4. How is astatine produced? What element is it most similar to?

____

____

5. Which two elements were discovered in the 1990s?

____

____

6. Name the rare Earth element that was discovered in 1789 and named for the planet Uranus. What is it used for?

____

____

____

Copyright © by Holt, Rinehart and Winston. All rights reserved.

Performance-Based Assessment

Teacher Notes and Answer Key

PURPOSE

Students produce an oxidation reaction and observe its effects. Students then use their knowledge of the periodic table to explain the reaction.

TIME REQUIRED

One 45-minute class period. Students will need 20 minutes to perform the procedure and 25 minutes to answer the analysis questions.

RATING

Teacher Prep–2
Student Set-Up–1
Concept Level–2
Clean Up–2

ADVANCE PREPARATION

Equip each activity station with the necessary materials. You may choose to copy the periodic table for students' use during this activity.

SAFETY INFORMATION

Use plastic bowls, if available. Use alcohol thermometers rather than mercury thermometers. Instruct students not to touch broken thermometers. Have a disposal container for sharps available in case of thermometer breakage. Perform this activity in a well-ventilated area. All students should wear safety goggles. Mop up spills immediately.

TEACHING STRATEGIES

This activity works best in groups of 2–3 students. The reaction of the iron in the steel wool with the oxygen in the air is classified as an oxidation-reduction (redox) reaction. The iron is the reducing agent that gives up its electrons. The oxygen in the air is the oxidizing agent that accepts electrons. The acid in vinegar facilitates the transfer of electrons by enhancing the conductivity of the moisture left on the steel wool. The two products of this reaction are iron (III) oxide (rust) and heat.

Copyright © by Holt, Rinehart and Winston. All rights reserved.

Performance-Based Assessment *continued*

Evaluation Strategies

Use the following rubric to help evaluate student performance.

Rubric for Assessment

Possible points	Appropriate use of materials and equipment (30 points possible)
30–20	Successfully completes activity; safe and careful handling of materials and equipment; attention to detail; superior lab skills
19–10	Task is generally complete; successful use of materials and equipment; sound knowledge of lab techniques; somewhat unfocused performance; mild neglect of safety measures
9–1	Attempts to complete tasks yield inadequate results; unsafe lab technique; apparent lack of skill
	Quality and clarity of observations (40 points possible)
40–30	Superior observations stated clearly and accurately; high level of detail; correct use of units of measurement
29–20	Accurate observations; moderate level of detail; correct use of units of measurement
19–10	Complete observations, but expressed in unclear manner; may include minor inaccuracies; attempts to use units of measurement include errors or inconsistencies
9–1	Erroneous, incomplete, or unclear observations; lack of accuracy, details, units of measurement
	Explanation of observations (30 points possible)
30–20	Clear, detailed explanation shows superior knowledge of the periodic table
19–10	Adequate understanding of periodic table concepts with minor difficulty in expression
9–1	Poor understanding of the periodic table; explanation unclear or not relevant to periodic table; substantial factual errors

Copyright © by Holt, Rinehart and Winston. All rights reserved.

Name _______________________________ Class _______________ Date _______________

Assessment

Performance-Based Assessment

OBJECTIVE

You've read about the periodic table. Now you will have a chance to observe the chemical properties of elements in action! In this activity you will cause a reaction between the iron in steel wool and the oxygen in the air.

KNOW THE SCORE!

As you work through the activity, keep in mind that you will be earning a grade for the following:

- how you work with materials and equipment (30%)
- the quality and clarity of your observations (40%)
- how you use the periodic table to explain those observations (30%)

Using Scientific Methods

ASK A QUESTION

What will happen when iron in steel wood reacts with the air?

MATERIALS AND EQUIPMENT

- graduated cylinder
- protective gloves
- vinegar
- small bowl

- fine steel wool, 0000 grade
- watch or clock
- thermometer
- rubber band

SAFETY INFORMATION 

- Do not touch broken thermometers.
- Wear safety goggles.

FORM A HYPOTHESIS

1. Describe the steel wool before the reaction.

   Sample answer: The steel wool looks like metallic cotton. The color is gray

   or dark silver.

TEST THE HYPOTHESIS

2. Use the graduated cylinder to measure 50 mL of vinegar. Pour the vinegar into the bowl.

3. Place the steel wool in the vinegar and leave it there for 2 minutes.

Copyright © by Holt, Rinehart and Winston. All rights reserved.

Name _______________________________ Class ________________ Date ______________

Performance-Based Assessment *continued*

4. With a gloved hand, remove the steel wool from the vinegar.

5. Wrap the steel wool around the thermometer. Use a rubber band to hold the steel wool in place.

6. Record the thermometer's starting temperature.

The starting temperature is 24°C.

7. Keep the steel wool around the thermometer for 10 minutes.

8. Record the thermometer's ending temperature.

The ending temperature is 30°C.

ANALYZE THE RESULTS

1. Describe the steel wool after the reaction in Step 3.

The parts of the steel wood that were soaked in vinegar appear rusty. The color changed from gray to orange.

2. Record the thermometer's starting temperature.

The starting temperature is 24°C.

3. Record the thermometer's ending temperature.

The ending temperature is 30°C.

4. List the two changes that you observed or measured after the reaction.

Rust appeared and the temperature of the steel wood rose by 6°C.

5. Use the locations of iron and oxygen in the periodic table to explain the change in the steel wool's appearance.

Iron is a reactive transition metal. Oxygen belongs to a group of reactive elements. The iron and oxygen reacted with one another, forming rust.

DRAW CONCLUSIONS

6. What would you expect to see if you put steel wool in a chamber filled with chlorine gas? Explain.

The steel wool would rust very quickly. Chlorine is more reactive than oxygen.

____

Copyright © by Holt, Rinehart and Winston. All rights reserved.

Name _________________________________ Class _________________ Date _____________

Performance-Based Assessment *continued*

7. What would you expect to see if you put the steel wool in a chamber filled with neon gas? Explain.

The steel wool and neon would not react. Neon is a noble gas and is

therefore nonreactive.

8. Iron is a transition metal. List two common properties of transition metals.

Answers will vary. Sample answer: good conductors of thermal energy and

electric current, high density, high melting point, less reactive than alkali

metals and alkaline-earth metals, silver-colored.

Copyright © by Holt, Rinehart and Winston. All rights reserved.

 Model Making)

DATASHEET FOR CHAPTER LAB

Create a Periodic Table

Norman Holcomb
Marion Elementary School
Maria Stein, Ohio

Teacher Notes and Answer Key

TIME REQUIRED

One to two 45-minutes class periods

RATING

Teacher Prep–3
Student Set-Up–1
Concept Level–3
Clean Up–1

MATERIALS

The materials listed for this lab are for each group of 2–4 students. For each group of students, assemble a collection of 20 objects (five sets of four objects). You should provide a bag containing 19 of these objects. A recommended collection of objects includes sets of coins (penny, nickel, dime, quarter), sets of buttons that are similar but vary in diameter, and washers that vary in diameter. Other objects, such as nuts, bolts, and paper circles, will work and are easily obtainable. The difference in masses should be large enough for a beam balance to detect. Ideally, each set (one column on the table) should be of the same material and thickness and vary only in diameter.

PREPARATION NOTE

You may have students prepare the 20 squares of paper, but the lab will go faster if the squares are prepared ahead of time.

Copyright © by Holt, Rinehart and Winston. All rights reserved.

Name _________________________________ Class _______________ Date ____________

Model-Making   **DATASHEET FOR CHAPTER LAB**

Create a Periodic Table

You probably have classification systems for many things in your life, such as your clothes, your books, and your CDs. One of the most important classification systems in science is the periodic table of the elements. In this lab, you will develop your own classification system for a collection of ordinary objects. You will analyze trends in your system and compare your system with the periodic table of the elements.

OBJECTIVES

Classify objects based on their properties.

Identify patterns and trends in data.

MATERIALS

- bag of objects
- balance, metric
- meterstick
- paper, graphing (2 sheets)
- paper, 3 × 3 cm squares (20)

PROCEDURE

1. Your teacher will give you a bag of objects. Your bag is missing one item. Examine the items carefully. Describe the missing object in as many ways as you can. Be sure to include the reasons why you think the missing object has the characteristics you describe.

   ____

   ____

   ____

2. Lay the paper squares out on your desk or table so that you have a grid of five rows of four squares each.

3. Arrange your objects on the grid in a logical order. (You must decide what order is logical!) You should end up with one blank square for the missing object.

4. Record a description of the basis for your arrangement.

   ____

5. Measure the mass (g) and diameter (mm) of each object, and record your results in the appropriate square. Each square (except the empty one) should have one object and two written measurements on it.

Copyright © by Holt, Rinehart and Winston. All rights reserved.

Name _______________________________ Class _______________ Date _____________

Create a Periodic Table *continued*

6. Examine your pattern again. Does the order in which your objects are arranged still make sense? Explain.

____

____

7. Rearrange the squares and their objects if necessary to improve your arrangement. Record a description of the basis for the new arrangement.

____

____

8. Working across the rows, number the squares 1 to 20. When you get to the end of a row, continue numbering in the first square of the next row.

9. Copy your grid below. In each square, be sure to list the type of object and label all measurements with appropriate units.

ANALYZE THE RESULTS

1. Constructing Graphs On a separate piece of paper make a graph of mass (y-axis) versus object number (x-axis). Label each axis, and title the graph.

2. Constructing Graphs Now make a graph of diameter (y-axis) versus object number (x-axis).

Copyright © by Holt, Rinehart and Winston. All rights reserved.

Name _________________________________ Class _______________ Date ____________

Create a Periodic Table *continued*

DRAW CONCLUSIONS

3. Analyzing Graphs Discuss each graph with your classmates. Try to identify any important features of the graph. For example, does the graph form a line or a curve? Is there anything unusual about the graph? What do these features tell you? Record your answers.

Answers will vary. The primary feature is the repeating pattern of increases. This pattern in the first graph indicates the periodic nature of the mass of the items. This pattern in the second graph indicates the periodic nature of the diameter of the items.

4. Evaluating Models How is your arrangement of objects similar to the periodic table of the elements found in this textbook? How is your arrangement different from that periodic table?

Answers will vary. Similarities include repeating patterns (such as mass) across the table. Differences may include no consistent family traits and no chemical properties associated with position in the table.

5. Making Predictions Look again at your prediction about the missing object. Do you think your prediction is still accurate? Try to improve your description by estimating the mass and diameter of the missing object. Record your estimates.

Answers will vary, depending on the student's original prediction. Accept all reasonable answers. (You may wish to provide the students with the missing object so they can further evaluate their prediction.)

6. Evaluating Methods Mendeleev created a periodic table of elements and predicted characteristics of missing elements. How is your experiment similar to Mendeleev's work?

This experiment is similar in that a pattern was identified that helped to identify characteristics of a missing object.

Copyright © by Holt, Rinehart and Winston. All rights reserved.

Name _______________________________ Class _______________ Date _____________

DATASHEET FOR QUICK LAB

Conduction Connection

MATERIALS

- cup, plastic-foam
- graphite, mechanical pencil lead
- water, hot
- wire, copper, bare

SAFETY INFORMATION 

PROCEDURE

1. Fill a **plastic-foam cup** with **hot water**.

2. Stand a **piece of copper wire** and a **graphite lead** from a mechanical pencil in the water.

3. After 1 min, touch the top of each object. Record your observations.

____

____

____

4. Which material conducted thermal energy the best? Why?

   The wire conducted thermal energy better than the pencil lead did. The wire

   is made of the metal copper; pencil lead is made of graphite, a form of the

   nonmetal carbon. Metals conduct thermal energy better than nonmetals do.

Copyright © by Holt, Rinehart and Winston. All rights reserved.

Answer Key

Directed Reading A

SECTION: ARRANGING THE ELEMENTS

1. Answers will vary. Sample answer: Scientists might have been frustrated because the elements weren't organized and therefore their properties couldn't be predicted.
2. D
3. periodic
4. periodic table of the elements
5. Mendeleev was able to predict the properties of unknown elements by using the pattern of properties in the periodic table.
6. D
7. periodic law
8. C
9. Chemical symbols are color-coded on the periodic table according to state. The color of the chemical symbol for carbon is red, which corresponds to a solid.
10. properties
11. electrons
12. Answers will very. Sample answer: The zigzag line can help me recognize which elements are metals, which are nonmetals, and which are metalloids.
13. metals
14. ductile
15. solid
16. It means that most metals can be flattened with a hammer and will not shatter.
17. aluminum
18. nonmetals
19. metalloids
20. B
21. C
22. chemical symbol
23. periods
24. groups, families
25. mendelevium, californium

SECTION: GROUPING THE ELEMENTS

1. C
2. B
3. alkali metals
4. alkaline-earth metals
5. cement and chalk
6. Answers will vary. Sample answer: Calcium is an important part of a compound that keeps your bones and teeth healthy.
7. Accept three from this list: beryllium, magnesium, strontium, barium, and radium.
8. B
9. transition metals
10. Answers will vary. Sample answer: Mercury is in a liquid state at room temperature. The other transition metals are solids at room temperature.
11. lanthanides, actinides
12. europium
13. americium
14. Answers will vary. Sample answer: Aluminum was considered more valuable than gold.
15. Answers will vary. Sample answer: Aluminum is used in making aircraft parts, lightweight automobile parts, foil, cans, and siding.
16. silicon, germanium
17. proteins, fats, and carbohydrates
18. diamond
19. Answers will vary. Sample answer: used as a jewel and on cutting tools, such as saws, drills, and files.
20. soot
21. gas
22. five
23. hydrogen
24. Answers will vary. Sample answer: Oxygen is a gas at room temperature. The other four elements are solids.
25. sulfur
26. Answers will vary. Sample answer: Oxygen, which makes up about 20% of air, is important to most living things. It is also necessary for substances to burn.
27. halogens
28. Both are used as disinfectants.
29. salts, sodium chloride

Copyright © by Holt, Rinehart and Winston. All rights reserved.

30. Answers will vary. Sample answer: Chlorinating water helps protect people from many diseases by killing the organisms that cause the diseases.
31. C
32. noble gases
33. density
34. D

Directed Reading B

SECTION: ARRANGING THE ELEMENTS

1. D
2. A
3. C
4. C
5. B
6. A
7. C
8. A
9. metals
10. nonmetals
11. metalloids
12. semiconductors
13. mendelevium
14. californium
15. chemical symbol
16. C
17. C
18. C
19. B
20. B

SECTION: GROUPING THE ELEMENTS

1. C
2. B
3. B
4. B
5. A
6. D
7. A
8. C
9. C
10. A
11. reactive
12. aluminum
13. carbohydrates
14. tin
15. hydrogen
16. oxygen
17. A
18. B
19. D
20. C
21. C
22. A
23. C

Vocabulary and Section Summary

SECTION: ARRANGING THE ELEMENTS

1. periodic: describes something that occurs or repeats at regular intervals
2. periodic law: the law that states that the repeating chemical and physical properties of elements change periodically with the atomic numbers of the elements
3. period: in chemistry, a horizontal row of elements in the periodic table
4. group: a vertical column of elements in the periodic table; elements in a group share chemical properties

SECTION: GROUPING THE ELEMENTS

1. alkali metal: one of the elements of Group 1 of the periodic table (lithium, sodium, potassium, rubidium, cesium, and francium)
2. alkaline-earth metal: one of the elements of Group 2 of the periodic table (beryllium, magnesium, calcium, strontium, barium, and radium)
3. halogen: one of the elements of Group 17 of the periodic table (fluorine, chlorine, bromine, iodine, and astatine); halogens combine with most metals to form salts
4. noble gas: one of the elements of Group 18 of the periodic table (helium, neon, argon, krypton, xenon, and radon); noble gases are unreactive

Section Review

SECTION: ARRANGING THE ELEMENTS

1. Sample answer: *Periodic* means "happening in a regular repeating pattern."
2. C
3. A period in the periodic table is a horizontal row of elements. A group is a vertical column of elements.
4. atomic mass
5. The repeating chemical and physical properties of elements change periodically with the atomic numbers of the elements.

Copyright © by Holt, Rinehart and Winston. All rights reserved.

6. halogen; Element 117 has 117 protons. So, it would fall under astatine in the periodic table.

7. lithium; Sodium and lithium are in the same group, so their properties should be more alike than the properties of sodium and magnesium are.

8. The periodic table has the same shape, atomic numbers, and chemical symbols. The names of the elements are in a different language (Japanese).

SECTION: GROUPING THE ELEMENTS:

1. noble gas

2. alkali metal

3. halogen

4. alkaline-earth metal

5. C

6. Answers may vary but could include that they have one electron in their outer level; are very reactive; are soft, silver-colored, and shiny; and have a low density.

7. having the same number of electrons in the outer level of their atoms

8. Answers may vary but could include that they have seven electrons in their outer level, are very reactive, conduct electric current poorly, react violently with alkali metals to form salts, and are never found uncombined in nature.

9. The properties of hydrogen do not match the properties of any single group.

10. boron group (Group 13)

11. metal; The model shows two electrons in the outer level, so the atom represented is most likely a metal.

12. They are so reactive that they react with water or oxygen in the air.

13. Both hydrogen and sodium have one electron in their outer level. Atoms of both elements give away one electron when joining with other atoms. However, hydrogen is a nonmetal and is a gas at room temperature, whereas sodium is a solid metal at room temperature.

Chapter Review

1. group

2. period

3. alkali metals

4. noble gases

5. D

6. B

7. C

8. C

9. B

10. C

11. Moseley arranged elements by increasing atomic number. Mendeleev arranged elements by increasing atomic mass.

12. Both are periodic. The periodic table has repeating properties of elements. The calendar has repeating days and months.

13. 5.4% (sodium and potassium)

14. 5.6% (magnesium and calcium)

15. An answer to this exercise can be found at the end of the teacher's edition.

16. Mendeleev could make predictions only about elements where there were clear gaps in his table. Because no noble gases were known at the time, there were no obvious gaps in the table and no way that he could have known that a whole column was missing.

17. metal; it will be located below the metal bismuth to the left of the zigzag

18. I would tell my classmate that he didn't find sodium. Sodium is very reactive and cannot be found uncombined in nature. Sodium would react with oxygen and water in the air and form a compound.

19. a. calcium
b. carbon

20. Periodic properties are the order of the shapes and the number of lines inside the shape. The properties shared in a group are the shape and the color of the lines inside the shape.

Copyright © by Holt, Rinehart and Winston. All rights reserved.

Reinforcement

PLACING ALL YOUR ELEMENTS ON THE TABLE

1.–13. Make sure that students have colored the table properly.

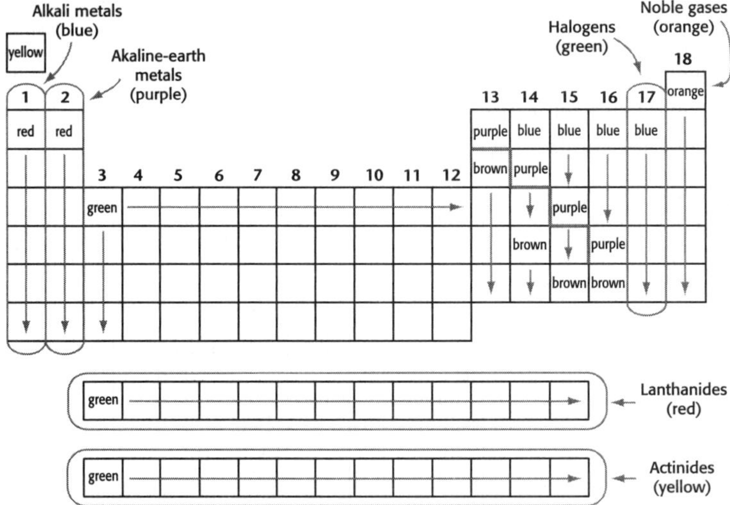

14. They are in Group 2.

15. 2

16. Answers will vary. Sample answer: The alkali metals are solids, while hydrogen is a gas at room temperature.

17. They are called the noble gases.

18. B

19. Group 13

20. C

21. Accept all answers.

Critical Thinking

1. Answers will vary. Sample answer: Sodium is not found by itself naturally. Because of its reactivity, sodium always combines with other elements.

2. Answers will vary. Sample answer: No; sulfur, a nonmetal, is a poor conductor of heat and energy. An electrical current could not travel along sulfur wire.

3. Answers will vary. Sample answer: Yes; an explosion would result. Hydrogen reacts explosively when combined with oxygen in this manner.

4. Answers will vary. Sample answer: No; the oxygen in the "Everlast Light Bulb" would react with the heating filament and cause it to burn out more quickly than a light bulb filled with a nonreactive gas, such as argon.

5. Answers will vary. Sample answer: If the substance could not be broken down any further and if its chemical and physical properties were not identical to any other element, the claim would be true.

6. Answers will vary. Sample answer: The element would be classified according to its chemical and physical properties.

Copyright © by Holt, Rinehart and Winston. All rights reserved.

Section Quizzes

SECTION: ARRANGING THE ELEMENTS

1. A
2. C
3. B
4. A
5. A

SECTION: GROUPING THE ELEMENTS

1. H
2. I
3. C
4. E
5. G
6. J
7. B
8. A
9. F
10. D

Chapter Test A

1. A
2. D
3. C
4. B
5. B
6. B
7. A
8. A
9. D
10. F
11. A
12. E
13. G
14. J
15. C
16. I
17. H
18. D
19. B
20. A
21. D
22. B
23. B
24. A

Chapter Test B

1. periodic
2. periodic law

3. halogens
4. alkali metals
5. alkaline-earth metals
6. group
7. C
8. A
9. A
10. D
11. A
12. A
13. Answers will vary. Sample answer: The periodic law states that chemical and physical properties of elements are periodic, repeating functions of the elements' atomic numbers. This is why elements in vertical groups of the periodic table share similar properties.
14. Answers will vary. Sample answer: The properties of hydrogen do not match the properties of any single group.
15. Answers will vary. Sample answer: Transition metals tend to be shiny and to conduct thermal energy and electric current well.
16. Answers will vary. Sample answer: The lanthanides and actinides are transition metals. The lanthanides are shiny, reactive metals. The actinides are radioactive.
17. Answers will vary. Sample answer: Life could not exist without carbon. Carbon forms compounds such as proteins, fats, and carbohydrates that are necessary for living things on Earth.
18. Answers will vary. Sample answer: When light bulbs are filled with argon, they last longer. Argon is unreactive and so does not react with the metal filament in a light bulb. The low density of helium makes blimps and weather balloons float.
19. **a.** metals; **b.** metalloids; **c.** nonmetals; **d.** solids; **e.** shiny; **f.** gases

Chapter Test C

1. A
2. D
3. A
4. D
5. C
6. A
7. B

Copyright © by Holt, Rinehart and Winston. All rights reserved.

8. A
9. C
10. C
11. A
12. B
13. D
13. D
14. D
15. A
16. C
17. B
18. aluminum
19. carbohydrates
20. hydrogen
21. silicon
22. oxygen

Standardized Test Preparation

READING

Passage 1
  1. A
  2. G
  3. B

Passage 2
  1. C
  2. F
  3. D

INTERPRETING GRAPHICS
  1. D
  2. F
  3. D

MATH
  1. D
  2. I
  3. C

Vocabulary Activity

  1. periodic law
  2. group
  3. transition metal

4. alkali metals
5. periodic
6. alkaline-earth
7. period
8. noble gases
9. F
10. Y
11. Xe
12. metalloids
13. lanthanides
14. hydrogen
15. halogens

Mendeleev's Quotation: The elements, if arranged according to their atomic weights, exhibit a periodicity of properties.

SciLinks Activity

  1. Answers will vary. Sample answer: Gold is a heavy, yellow, metallic chemical element. It is a precious metal with a high degree of ductility and malleability. It is used in the manufacture of jewelry, electronics, and coins.
  2. the troy ounce
  3. Answers will vary. Sample answer: Gold circuits ar used in the electronic sensors that operate airbags. These sensors ensure that airbags operate without failing, saving many lives each year.
  4. Astatine is produced by bombarding bismuth with alpha particles. It is most similar to iodine.
  5. Ununnilium (Uuu) was discovered in 1994. Darmstatium (Ds), which was temporarily called Ununbium (Uub) was discovered in 1996.
  6. Uranium was discovered in 1789 and named for the planet Uranus. It is used as fuel for nuclear reactors.

Copyright © by Holt, Rinehart and Winston. All rights reserved.

Lesson Plan

Section: Arranging the Elements

Pacing

Regular Schedule: **with lab(s):** 2 days   **without lab(s):** 1 day

Block Schedule: **with lab(s):** 1 day   **without lab(s):** 0.5 day

Objectives

1. Describe how Mendeleev arranged elements in the first periodic table.

2. Explain how elements are arranged in the modern periodic table.

3. Compare metals, nonmetals, and metalloids based on their properties and on their location in the periodic table.

4. Describe the difference between a period and a group.

National Science Education Standards Covered

UCP 1: Systems, order, and organization

SAI 2: Understandings about scientific inquiry

SPSP 5: Science and technology in society

HNS 1: Science as a human endeavor

HNS 2: Nature of science

HNS 3: History of science

PS 1b: Substances react chemically in characteristic ways with other substances to form new substances (compounds) with different characteristic properties. In chemical reactions, the total mass is conserved. Substances often are placed in categories or groups if they react in similar ways; metals is an example of such a group.

KEY

SE = Student Edition   **TE** = Teacher's Edition

CRF = Chapter Resource File

FOCUS (5 minutes)

- **Chapter Starter Transparency**   Use this transparency to introduce the chapter.

- **Bellringer, TE**   Have students determine patterns that can be established with a deck of cards.

- **Bellringer Transparency**   Use this transparency as students enter the classroom and find their seats.

Copyright © by Holt, Rinehart and Winston. All rights reserved.

_ **Reading Strategy, SE** Have students create a mnemonic device to remember the difference between groups and periods.

MOTIVATE *(10 minutes)*

_ **Demonstration, Grouping, TE** Ask the class to group student volunteers based on lists of characteristics. **(GENERAL)**

TEACH *(65 minutes)*

_ **Activity, Element Sampling, TE** Have students identify and list characteristics of samples of various elements. **(GENERAL)**

_ **Using the Figure, TE** Tell students that the color scheme on the periodic table will be continued throughout the book. **(GENERAL)**

_ **Activity, Looking for Gaps, TE** Have students consider the progression of atomic mass across the periodic table. **(ADVANCED)**

_ **Research, TE** Have students find out if any new elements have been synthesized since this book has been published. **(GENERAL)**

_ **Connection to History, TE** Tell students about the early steps that led to the creation of the modern periodic table. **(GENERAL)**

_ **Quick Lab, Conduction Connection, SE** Students determine whether graphite or copper conducts thermal energy the best. **(GENERAL)**

_ **Activity, Element Game, TE** Have students make a memory game to practice learning element names and symbols. **(BASIC)**

_ **Directed Reading A/B, CRF** These worksheets reinforce basic concepts and vocabulary presented in the lesson. **(Basic/Special Needs)**

_ **Vocabulary and Section Summary, CRF** Students write definitions of key terms and read a summary of section content. **(GENERAL)**

_ **Teaching Transparency, The Periodic Table of the Elements** Use this graphic to help students understand the organization of the periodic table.

_ **Chapter Lab, Create a Periodic Table, SE** Students create a periodic table using ordinary objects. **(GENERAL)**

_ **Datasheet for Chapter Lab, Create a Periodic Table, CRF** Students use the datasheet to complete the Chapter Lab. **(GENERAL)**

CLOSE *(10 minutes)*

_ **Reteaching, TE** Reinforce the meanings of the terms malleable and brittle.

_ **Section Review, SE** Students answer end-of-section vocabulary, key ideas, critical thinking, and interpreting graphics questions. **(GENERAL)**

_ **Section Quiz, CRF** Students answer five objective questions about the elements on the periodic table. **(GENERAL)**

_ **Quiz, TE** Students answer three questions about the elements on the periodic table. **(GENERAL)**

Copyright © by Holt, Rinehart and Winston. All rights reserved.

Lesson Plan

Section: Grouping the Elements

Pacing

Regular Schedule: **with lab(s):** N/A   **without lab(s):** 1 days

Block Schedule: **with lab(s):** N/A   **without lab(s):** 0.5 day

Objectives

1. Explain why elements in a group often have similar properties.

2. Describe the properties of the elements in the groups of the periodic table.

National Science Education Standards Covered

ST 2: Understandings about science and technology

PS 1b: Substances react chemically in characteristic ways with other substances to form new substances (compounds) with different characteristic properties. In chemical reactions, the total mass is conserved. Substances often are placed in categories or groups if they react in similar ways; metals is an example of such a group.

PS 3e: In most chemical and nuclear reactions, energy is transferred into or out of a system. Heat, light, mechanical motion, or electricity might all be involved in such transfers.

KEY
SE = Student Edition   **TE** = Teacher's Edition
CRF = Chapter Resource File

FOCUS *(5 minutes)*

_ **Bellringer, TE** Have students determine the characteristics of several animals and tell how determining the characteristics of elements can help them tell the elements apart.

_ **Bellringer Transparency** Use this transparency as students enter the classroom and find their seats.

MOTIVATE *(10 minutes)*

_ **Discussion, TE** Discuss cookie ingredients, comparing them with ingredients for the universe—the periodic table of the elements. (**GENERAL**)

Copyright © by Holt, Rinehart and Winston. All rights reserved.

TEACH *(20 minutes)*

_ **Reading Strategy, SE** Have students read the section silently and, in pairs, summarize what they've read.

_ **Using the Figure, Group Trends, TE** Have students describe how lithium should react with water.

_ **Cultural Awareness, TE** Explain the Arabic origins of alkali metals. **(GENERAL)**

_ **Connection to Life Science, TE** Explain the importance of calcium to health. **(GENERAL)**

_ **Connection to History, Canning, TE** Describe the development of food preservation through canning. **(GENERAL)**

_ **Using the Figure, Allotropes, TE** Explain allotropes to students. **(GENERAL)**

_ **Connection Activity History, Sulfur, TE** Have students write a report, make a poster, or prepare a presentation on the uses of sulfur prior to 1777. **(ADVANCED)**

_ **Cultural Awareness, The Curies, TE** Tell students about the accomplishments of the Curies. **(GENERAL)**

_ **Connection to Astronomy, TE** Have students research how elements may be created or changed in violent celestial reactions. **(GENERAL)**

_ **Directed Reading A/B, CRF** These worksheets reinforce basic concepts and vocabulary presented in the lesson. **(Basic/Special Needs)**

_ **Vocabulary and Section Summary, CRF** Students write definitions of key terms and read a summary of section content. **(GENERAL)**

_ **Reinforcement, CRF** This worksheet reinforces key concepts in the chapter. **(BASIC)**

_ **Critical Thinking, CRF** Ask students to fill out the worksheet about new science products found on an advertisement. **(ADVANCED)**

_ **SciLinks Activity, The Periodic Table, SciLinks code HSM1125, CRF** Students research Internet resources related to gold. **(GENERAL)**

CLOSE *(10 minutes)*

_ **Reteaching, TE** Teach students a game to help them remember the properties of the groups on the periodic table. **(GENERAL)**

_ **Section Review, SE** Students answer end-of-section vocabulary, key ideas, interpreting graphics, and critical thinking questions. **(GENERAL)**

_ **Section Quiz, CRF** Students answer ten objective questions about the elements on the periodic table. **(GENERAL)**

_ **Quiz, TE** Students answer three questions about the elements on the periodic table. **(GENERAL)**

_ **Alternative Assessment, TE** Students will prepare a concept map of the periodic table. **(GENERAL)**

Copyright © by Holt, Rinehart and Winston. All rights reserved.

Lesson Plan

End of Chapter Review and Assessment

Pacing

Regular Schedule: **with lab(s):** N/A   **without lab(s):** 1 day

Block Schedule: **with lab(s):** N/A   **without lab(s):** 0.5 day

---

KEY
SE = Student Edition   **TE** = Teacher's Edition
CRF = Chapter Resource File

_ **Chapter Review, SE** Students answer end-of-chapter vocabulary, key ideas, critical thinking, and graphics questions. (**GENERAL**)

_ **Vocabulary Activity, CRF** Students review chapter vocabulary terms by completing a quotation puzzle. (**GENERAL**)

_ **Chapter Test A/B/C, CRF** Assign questions from the appropriate test for chapter assessment. (**General/Advanced/Special Needs**)

_ **Performance-Based Assessment, CRF** Assign this activity for general level assessment for the chapter. (**GENERAL**)

_ **Standardized Test Preparation, CRF** Students answer reading comprehension, math, and interpreting graphics questions in the format of a standardized test. (**GENERAL**)

_ **Test Generator, One-Step Planner** Create a customized homework assignment, quiz, or test using HRW Test Generator program.

Copyright © by Holt, Rinehart and Winston. All rights reserved.

The Periodic Table

MULTIPLE CHOICE

1. *Periodic* means
   a. happening at regular intervals.
   b. happening very rarely.
   c. happening frequently.
   d. happening three or four times a year.
   Answer: A   Difficulty: 1   Section: 1   Objective: 1

2. Periodic law states that
   a. elements are either gases, solids, or liquids.
   b. mercury is a liquid at room temperature.
   c. properties of elements change periodically with the elements' atomic numbers.
   d. some elements only stay in a liquid state for short periods.
   Answer: C   Difficulty: 1   Section: 1   Objective: 2

3. Each vertical column on the periodic table is called a(an)
   a. period.
   b. group.
   c. element.
   d. property.
   Answer: B   Difficulty: 1   Section: 1   Objective: 4

4. The elements to the right of the zigzag line on the period table are called
   a. nonmetals.
   b. metals.
   c. metalloids.
   d. conductors.
   Answer: A   Difficulty: 1   Section: 1   Objective: 3

5. Most metals are
   a. solid at room temperature.
   b. bad conductors of electric current.
   c. dull.
   d. not malleable.
   Answer: A   Difficulty: 1   Section: 1   Objective: 3

6. Most of the elements in the periodic table are
   a. metals.
   b. metalloids.
   c. gases.
   d. nonmetals.
   Answer: A   Difficulty: 1   Section: 1   Objective: 3

7. Mendeleev arranged the elements by
   a. density.
   b. melting point.
   c. appearance.
   d. increasing atomic mass.
   Answer: D   Difficulty: 1   Section: 1   Objective: 1

8. The horizontal row on the period table is called a(n)
   a. group.
   b. family.
   c. period.
   d. atomic number.
   Answer: C   Difficulty: 1   Section: 1   Objective: 4

9. Which one of the following tells the physical state of an element at room temperature?
   a. the atomic number
   b. the color of the chemical symbol
   c. the atomic mass
   d. the element name
   Answer: B   Difficulty: 1   Section: 1   Objective: 2

10. How do the physical and chemical properties of the elements change?
    a. within a group
    b. across each period
    c. within a family
    d. across each group
    Answer: B   Difficulty: 1   Section: 1   Objective: 2

Copyright © by Holt, Rinehart and Winston. All rights reserved.

11. What is necessary for substances to burn?
    a. hydrogen
    b. oxygen
    c. helium
    d. carbon
    Answer: B   Difficulty: 1   Section: 2   Objective: 2

12. Transition metals are
    a. good conductors of thermal energy.
    b. more reactive than alkali metals.
    c. not good conductors of electric current.
    d. used to make aluminum.
    Answer: A   Difficulty: 1   Section: 2   Objective: 2

13. The elements' properties follow a pattern that repeats every
    a. 7 elements.
    b. 5 elements.
    c. 14 elements.
    d. 10 elements.
    Answer: A   Difficulty: 1   Section: 1   Objective: 1

14. The vertical column of elements on the periodic table is called a(n)
    a. period.
    b. semiconductor.
    c. atomic mass.
    d. group.
    Answer: D   Difficulty: 1   Section: 1   Objective: 4

15. What element makes up about 20% of the air we breathe?
    a. nitrogen
    b. bromine
    c. oxygen
    d. sulfur
    Answer: C   Difficulty: 1   Section: 2   Objective: 2

16. Mendeleev found that the elements' properties followed a pattern that repeated every
    a. 7 elements.
    b. 5 elements.
    c. 14 elements.
    d. 10 elements.
    Answer: A   Difficulty: 1   Section: 1   Objective: 1

17. The groups of elements that do not have individual names are called the
    a. transition metals.
    b. alkali metals.
    c. alkaline-earth metals.
    d. nonmetals.
    Answer: A   Difficulty: 1   Section: 2   Objective: 2

18. The carbon group has two metalloids, both of which are used to make
    a. dinnerware.
    b. foil.
    c. cans.
    d. computer chips.
    Answer: D   Difficulty: 1   Section: 2   Objective: 2

19. Diamond and soot are very different, yet both are natural forms of
    a. carbon.
    b. nickel.
    c. boron.
    d. copper.
    Answer: A   Difficulty: 1   Section: 2   Objective: 2

20. What element is used to make the most widely used compound in the chemical industry?
    a. sulfur
    b. tellurium
    c. selenium
    d. polonium
    Answer: A   Difficulty: 1   Section: 2   Objective: 2

21. What are most of the elements in the periodic table?
    a. metals
    b. metalloids
    c. precious metals
    d. nonmentals
    Answer: A   Difficulty: 1   Section: 1   Objective: 3

22. How did Mendeleev group the elements?
    a. by density
    b. by melting point
    c. by appearance
    d. by increasing atomic mass
    Answer: D   Difficulty: 1   Section: 1   Objective: 1

Copyright © by Holt, Rinehart and Winston. All rights reserved.

23. Mendeleev's pattern repeated after how many elements?
    a. every seven elements
    b. every three elements
    c. every five elements
    d. every two elements
    Answer: A   Difficulty: 1   Section: 1   Objective: 1

24. How would you describe most metals?
    a. They are easily shattered.
    b. They are bad conductors of electric current.
    c. They are dull.
    d. They can be drawn into thin wires.
    Answer: D   Difficulty: 1   Section: 1   Objective: 3

25. How many of the recently discovered elements follow periodic law?
    a. none of them
    b. every eighth element
    c. all of them
    d. half of them
    Answer: C   Difficulty: 1   Section: 1   Objective: 2

26. What are the left-to-right rows on the periodic table?
    a. periods
    b. families
    c. properties
    d. groups
    Answer: A   Difficulty: 1   Section: 1   Objective: 4

27. Which of the following is a property of alkali metals?
    a. They are so hard they cannot be cut.
    b. They are very reactive.
    c. They are stored in water.
    d. They have few uses.
    Answer: B   Difficulty: 1   Section: 2   Objective: 2

28. When a halogen reacts with a metal, what is formed?
    a. a salt
    b. a compound
    c. a nonmetal
    d. an electron
    Answer: A   Difficulty: 1   Section: 2   Objective: 2

29. What helps light bulbs last longer?
    a. krypton
    b. xenon
    c. argon
    d. neon
    Answer: C   Difficulty: 1   Section: 2   Objective: 2

COMPLETION

Use the terms from the following list to complete the sentences below.

    halogens   periodic law
    period   periodic
    group   noble gases
    alkali metals   alkaline-earth metals
    actinide

30. The days of the week are ___________________ because they repeat in the same
    order every seven days.
    Answer: periodic   Difficulty: 1   Section: 1   Objective: 1

31. A rule that states that repeating chemical and physical properties of elements change
    periodically with the atomic number of the elements is the ___________________.
    Answer: periodic law Difficulty: 1   Section: 1   Objective: 2

32. Iodine and chlorine are examples of ___________________.
    Answer: halogens   Difficulty: 1   Section: 2   Objective: 2

Copyright © by Holt, Rinehart and Winston. All rights reserved.

33. Pure ___________________ are often stored in oil to keep them from reacting with water and oxygen.
    Answer: alkali metals
    Difficulty: 1   Section: 2   Objective: 2

34. Atoms of ___________________ have two outer-level electrons.
    Answer: alkaline-earth metals
    Difficulty: 1   Section: 2   Objective: 2

35. Each up-and-down column of elements on the periodic table is called a(n) ___________________.
    Answer: group   Difficulty: 1   Section: 1   Objective: 4

Use the terms from the following list to complete the sentences below.

    aluminum   silicon
    oxygen   carbohydrates
    hydrogen

36. In the Boron Group, the most common element is ___________________.
    Answer: aluminum   Difficulty: 1   Section: 2   Objective: 1

37. Proteins, fats, and ___________________, which are compounds of carbon, are necessary for life on Earth.
    Answer:
      carbohydrates
    Difficulty: 1   Section: 2   Objective: 2

38. Nitrogen and ___________________ can be combined to make ammonia.
    Answer: hydrogen   Difficulty: 1   Section: 2   Objective: 2

39. Germanium and ___________________ are used to make computer chips.
    Answer: silicon   Difficulty: 1   Section: 2   Objective: 2

40. A substance needs ___________________ to burn.
    Answer: oxygen   Difficulty: 1   Section: 2   Objective: 2

Use the terms from the following list to complete the sentences below.

    elements   periodic
    periodic law   group
    period   alkali
    transition

41. When something is ___________________, it occurs or repeats at regular intervals.
    Answer: periodic   Difficulty: 1   Section: 1   Objective: 1

42. Mendeleev's table became known as the periodic table of the ___________________.
    Answer: elements   Difficulty: 1   Section: 1   Objective: 1

43. All of the more than 30 elements discovered since 1914 follow the ___________________.
    Answer: periodic law Difficulty: 1   Section: 1   Objective: 2

44. Properties such as conductivity and reactivity change gradually from left to right in each ___________________.
    Answer: period   Difficulty: 1   Section: 1   Objective: 4

45. A family is also called a ___________________.
    Answer: group   Difficulty: 1   Section: 1   Objective: 4

Copyright © by Holt, Rinehart and Winston. All rights reserved.

46. Because they are so reactive, _____________________ metals are found only combined with other elements in nature.
    Answer: alkali   Difficulty: 1   Section: 2   Objective: 2

47. The elements in Groups 3–12 are known as _____________________ metals.
    Answer: transition   Difficulty: 1   Section: 2   Objective: 1

Use the terms from the following list to complete the sentences below.

alkali	transition
lanthanides	alkaline-earth
metalloidshalogens	
actinides	

48. Sodium and potassium are _____________________ metals.
    Answer: alkali   Difficulty: 1   Section: 2   Objective: 1

49. Some of the _____________________, the shiny, reactive transition metals, are used to make steel.
    Answer: lanthanides Difficulty: 1   Section: 2   Objective: 2

50. All the atoms of _____________________ are unstable.
    Answer: actinides   Difficulty: 1   Section: 2   Objective: 2

51. Chlorine and bromine are examples of _____________________.
    Answer: halogens   Difficulty: 1   Section: 2   Objective: 1

52. Semiconductors, also known as _____________________, have some properties of metals and some properties of nonmetals.
    Answer: metalloids   Difficulty: 1   Section: 1   Objective: 3

53. The _____________________ metals are shiny and have one or two electrons in their outer level.
    Answer: transition   Difficulty: 1   Section: 2   Objective: 2

54. Magnesium, calcium, and barium are _____________________ metals.
    Answer:
        alkaline-earth
                                 Difficulty: 1   Section: 2   Objective: 1

SHORT ANSWER

55. State the periodic law, which is the basis for the periodic table.
    Answer:
        Answers will vary. Sample answer: The periodic law states that chemical and physical properties of elements are periodic, repeating functions of the elements' atomic numbers. This is why elements in vertical groups of the periodic table share similar properties.
    Difficulty: 1   Section: 1   Objective: 2

56. Explain why hydrogen is unique.
    Answer:
        Answers will vary. Sample answer: The properties of hydrogen do not match the properties of any single group.
    Difficulty: 1   Section: 2   Objective: 2

Copyright © by Holt, Rinehart and Winston. All rights reserved.

57. What generalizations can you make about transition metals?

    Answer:
    Answers will vary. Sample answer: Transition metals tend to be shiny and to conduct
    thermal energy and electric current well.

    Difficulty: 1   Section: 2   Objective: 1

58. Who was Henry Moseley and what did he determine?

    Answer:
    Answers will vary. Sample answer: He was a British scientist who determined the
    number of protons—the atomic number—in an atom.

    Difficulty: 1   Section: 1   Objective: 2

59. What is periodic law?

    Answer:
    Answers will vary. Sample answer: Periodic law states that the repeating chemical and
    physical properties of elements change periodically with the atomic numbers of the
    elements.

    Difficulty: 1   Section: 1   Objective: 2

60. What does it mean when a chemical symbol is red?

    Answer:
    It means that the element is a solid at room temperature.

    Difficulty: 1   Section: 1   Objective: 2

61. Each element has two numbers. What is the top number? What is the bottom number?

    Answer:
    Top: atomic number; bottom: atomic mass

    Difficulty: 1   Section: 1   Objective: 2

62. What are three characteristics of metals?

    Answer:
    Answers will vary. Sample answer. Most metals are solid at room temperature, have
    few electrons in their outer energy level, and are shiny. Other possible answers: Most
    metals are ductile, good conductors of electric current and thermal energy, and are
    malleable.

    Difficulty: 1   Section: 1   Objective: 3

63. How would you describe boron?

    Answer:
    Answers will vary. Sample answer: Boron is a metalloid that is almost as hard as a
    diamond, but is also very brittle. At high temperatures, it is a good conductor of
    electric current.

    Difficulty: 1   Section: 1   Objective: 3

64. What are three characteristics of nonmetals?

    Answer:
    Answers will vary. Sample answer: Atoms of nonmetals have an almost complete set
    of electrons in their outer level. Nonmetals are not malleable or ductile. Other possible
    answers: Nonmetals are not shiny and are poor conductors of thermal energy and
    electric current.

    Difficulty: 2   Section: 1   Objective: 3

65. What is sodium chloride and what is it used for?

    Answer:
    Sodium chloride is table salt. It is used to flavor food.

    Difficulty: 1   Section: 2   Objective: 2

Copyright © by Holt, Rinehart and Winston. All rights reserved.

66. What are three alkaline-earth metals?

    Answer:
    Answers will vary. Sample answer: beryllium, magnesium, and calcium. Other possible answers: strontium, barium, and radium.

    Difficulty: 1   Section: 2   Objective: 1

67. How would you describe the elements that are listed after plutonium, which is element 94?

    Answer:
    Answers will vary. Sample answer: These elements do not occur in nature. They are made in laboratories. Other possible answers: They are radioactive, or unstable, and their atoms can change into atoms of another element.

    Difficulty: 1   Section: 2   Objective: 2

68. What does the periodic law state?

    Answer:
    The chemical and physical properties of elements are periodic functions of their atomic numbers.

    Difficulty: 1   Section: 1   Objective: 2

69. Using the periodic table, which elements are in the same group as oxygen?

    Answer:
    sulfur, selenium, tellurium, and polonium

    Difficulty: 1   Section: 1   Objective: 2

70. List five elements whose symbols don't seem to come from their English names; for example, Fe is iron.

    Answer:
    Others include K—potassium, Na—sodium, W—tungsten, Cu—copper, Ag—silver, and Au—gold.

    Difficulty: 1   Section: 1   Objective: 4

71. Using the periodic table, determine which two groups have highly reactive metals.

    Answer:
    on the left, Groups 1 and 2

    Difficulty: 2   Section: 2   Objective: 1

72. What are the actinides? What is one characteristic of all actinides?

    Answer:
    the elements that follow actinium with atomic numbers 90–103; they are all radioactive

    Difficulty: 1   Section: 2   Objective: 2

73. Of the gases oxygen, argon, chlorine, and neon, which would be the two most chemically reactive?

    Answer:
    oxygen and chlorine; argon and neon are in Group 18, the noble gases, which are very unreactive

    Difficulty: 3   Section: 2   Objective: 2

Copyright © by Holt, Rinehart and Winston. All rights reserved.

MATCHING

a. aluminum
b. transition metals
c. lanthanides
d. oxygen
e. actinides

f. carbon
g. halogens
h. alkali metals
i. alkaline-earth metals
j. noble gases

74. ___ metals that are so reactive that in nature they are found only combined with other elements

Answer: H   Difficulty: 1   Section: 2   Objective: 2

75. ___ metals that have two outer-level electrons

Answer: I   Difficulty: 1   Section: 2   Objective: 2

76. ___ shiny, reactive metals, some of which are used to make steel

Answer: C   Difficulty: 1   Section: 2   Objective: 2

77. ___ elements whose atoms are radioactive

Answer: E   Difficulty: 1   Section: 2   Objective: 2

78. ___ very reactive nonmetals

Answer: G   Difficulty: 1   Section: 2   Objective: 2

79. ___ unreactive nonmetals that do not react with other elements under normal conditions

Answer: J   Difficulty: 1   Section: 2   Objective: 2

80. ___ metals in Groups 3–12 that do not give away their electrons as easily as atoms of Groups 1 and 2

Answer: B   Difficulty: 1   Section: 2   Objective: 1

81. ___ the most common element in Group 13, the Boron Group

Answer: A   Difficulty: 1   Section: 2   Objective: 1

82. ___ nonmetal that forms a wide variety of compounds, such as proteins, fats, and carbohydrates

Answer: F   Difficulty: 1   Section: 2   Objective: 2

83. ___ an element that is necessary for substances to burn

Answer: D   Difficulty: 1   Section: 2   Objective: 2

a. aluminum
b. argon
c. halogens
d. nitrogen
e. oxygen

f. alkali metals
g. calcium
h. carbon
i. hydrogen
j. lanthanides

84. ___ These metals react with water to form hydrogen.

Answer: F   Difficulty: 1   Section: 2   Objective: 2

85. ___ This metal, part of the Boron Group, is used for aircraft parts.

Answer: A   Difficulty: 1   Section: 2   Objective: 2

86. ___ This is important to most living things.

Answer: E   Difficulty: 1   Section: 2   Objective: 2

87. ___ Cement and chalk are compounds of this.

Answer: G   Difficulty: 1   Section: 2   Objective: 2

88. ___ Some of these reactive metals are used to make steel.

Answer: J   Difficulty: 1   Section: 2   Objective: 2

89. ___ Chlorine and iodine are these.

Answer: C   Difficulty: 1   Section: 2   Objective: 2

90. ___ This is a colorless, odorless gas.

Answer: I   Difficulty: 1   Section: 2   Objective: 2

Copyright © by Holt, Rinehart and Winston. All rights reserved.

91. ___ Diamond and soot are forms of this.
   Answer: H   Difficulty: 1   Section: 2   Objective: 2
92. ___ This makes up about 80% of the air we breathe.
   Answer: D   Difficulty: 1   Section: 2   Objective: 2
93. ___ Light bulbs last longer when they are filled with this gas.
   Answer: B   Difficulty: 1   Section: 2   Objective: 2

   a. semiconductor   c. noble gases
   b. mendelevium   d. californium

94. ___ group made up of six nonmetals
   Answer: C   Difficulty: 1   Section: 2   Objective: 2
95. ___ another name for metalloid
   Answer: A   Difficulty: 1   Section: 1   Objective: 2
96. ___ element named after a scientist
   Answer: B   Difficulty: 1   Section: 1   Objective: 2
97. ___ element named after a state
   Answer: D   Difficulty: 1   Section: 1   Objective: 2

   a. radioactive   c. colorless
   b. shiny   d. shiny

98. ___ word meaning "group"
   Answer: C   Difficulty: 1   Section: 1   Objective: 4

99. ___ word describing actinides
   Answer: A   Difficulty: 1   Section: 2   Objective: 2

100. ___ word describing hydrogen
   Answer: C   Difficulty: 1   Section: 2   Objective: 2

101. ___ word describing most transition metals
   Answer: B   Difficulty: 1   Section: 2   Objective: 2

ESSAY

102. Compare the lanthanides and the actinides.
   Answer:
   Answers will vary. Sample answer: The lanthanides and actinides are transition
   metals. The lanthanides are shiny, reactive metals. The actinides are radioactive.
   Difficulty: 2   Section: 2   Objective: 2

103. Carbon forms many important compounds. Could life exist without carbon? Explain
   your answer.
   Answer:
   Answers will vary. Sample answer: Life could not exist without carbon. Carbon forms
   compounds such as proteins, fats, and carbohydrates that are necessary for living
   things on Earth.
   Difficulty: 3   Section: 2   Objective: 2

104. How does the unreactivity of noble gases make them useful?
   Answer:
   Answers will vary. Sample answer: When light bulbs are filled with argon, they last
   longer. Argon is unreactive and so does not react with the metal filament in a light
   bulb. The low density of helium makes blimps and weather balloons float.
   Difficulty: 2   Section: 2   Objective: 2

Copyright © by Holt, Rinehart and Winston. All rights reserved.

INTERPRETING GRAPHICS

Use the figure below to answer the following questions.

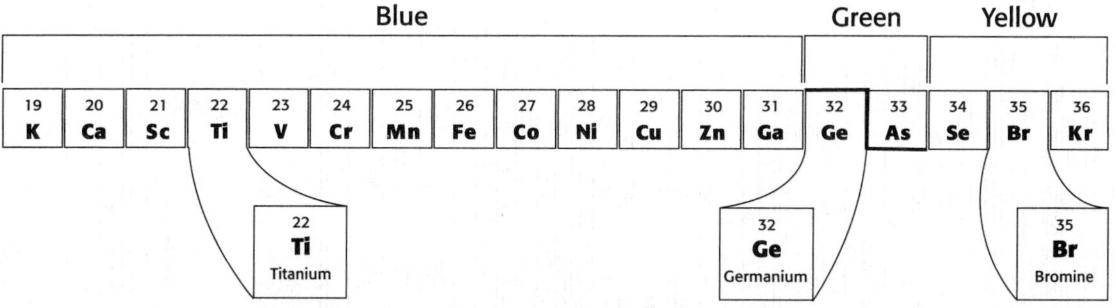

105. Which of the following elements is the most metallic?
    a. K
    b. Kr
    c. Fe
    d. Cu
    Answer: A   Difficulty: 1   Section: 1   Objective: 2

106. Which of these elements is the least metallic?
    a. V
    b. Zn
    c. Co
    d. Se
    Answer: D   Difficulty: 1   Section: 1   Objective: 2

107. Which element group has all nonmetals?
    a. K, Ca, Sc
    b. Se, Br, Kr
    c. V, Cr, Mn
    d. As, Se, Br
    Answer: B   Difficulty: 1   Section: 1   Objective: 3

Refer to the figure below to answer the following questions.

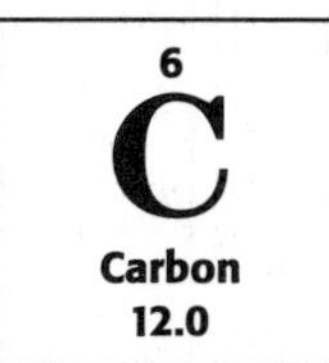

108. The number beneath carbon indicates the
    a. atomic number.
    b. atomic mass.
    c. chemical symbol.
    d. element name.
    Answer: B   Difficulty: 1   Section: 1   Objective: 2

109. The number at the top is the
    a. atomic number.
    b. element name.
    c. atomic mass.
    d. chemical symbol.
    Answer: A   Difficulty: 1   Section: 1   Objective: 2

Copyright © by Holt, Rinehart and Winston. All rights reserved.

CONCEPT MAPPING

110. Use the following terms to complete the concept map below:

nonmetals	metals
solids	gases
shiny	metalloids

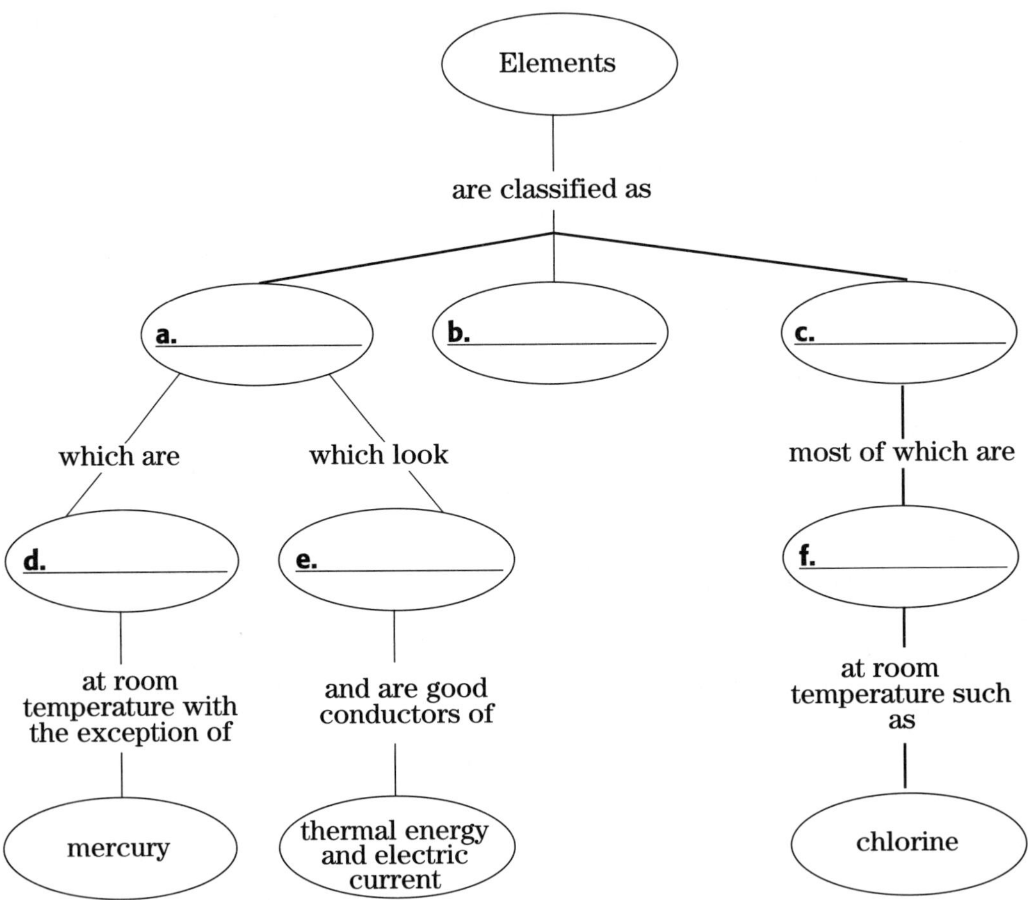

Answer:
a. metals; b. metalloids; c. nonmetals; d. solids; e. shiny; f. gases
Difficulty: 3   Section: 1   Objective: 2

Copyright © by Holt, Rinehart and Winston. All rights reserved.

9997282299   1 2 3 4 5 6